# U.S. ENERGY POLICY

Alternatives for Security

**DOUGLAS R. BOHI**
**MILTON RUSSELL**

Published for Resources for the Future, Inc.
by The Johns Hopkins University Press
Baltimore and London

77644

Resources for the Future is a nonprofit corporation for research and education in the development, conservation, and use of natural resources and the improvement of the quality of the environment. It was established in 1952 with the cooperation of the Ford Foundation. Part of the work of Resources for the Future is carried out by its resident staff; part is supported by grants to universities and other nonprofit organizations. Unless otherwise stated, interpretations and conclusions in RFF publications are those of the authors; the organization takes responsibility for the selection of significant subjects for study, the competence of the researchers, and their freedom of inquiry.

Douglas Bohi is associate professor of economics at Southern Illinois University, Carbondale, Illinois. Milton Russell is professor of economics at Southern Illinois University and is currently on leave as senior staff economist, Council of Economic Advisers. Charts were drawn by Federal Graphics. The book was edited by Ruth Haas.

*RFF editors:* Mark Reinsberg, Joan R. Tron, Ruth B. Haas, Joann Hinkel

Library of Congress Catalog Card Number 75-4209
ISBN 0-8018-1732-3      Paper $5.00

# Table of Contents

Preface      vii
Acknowledgments      ix
1.   OVERVIEW      1
     The Energy Shortage and Price Escalation   2
     Routes to Energy Security   5
     Summary of Results   9

2.   FUTURE IMPORT REQUIREMENTS AND
     DOMESTIC PRICES      14
     U.S. Supply–Demand Balance: The Base Case   15
     Sources of Projected Imports   21
     U.S. Imports at Higher Prices   21
     Import Controls versus Subsidies and Consumption
     Taxes   27

3.   POLICIES OF OIL-EXPORTING COUNTRIES AND
     U.S. ENERGY SECURITY      30
     Control of Oil Production and Pricing   31
       The Changing Role of the Majors   32
       Ascending Power of the Oil-Exporting Countries   34
       OPEC after the October War   37
     OPEC as a Producer Cartel   38
     The Long-Run Cartel Price   41
     Variations from the Long-Run Cartel Price   45
       Member Country Discount Rates   46

iii

Country Postures in Relation to the OPEC Price   52
The Lower Bound of the Cartel Price   54
  Determining the Country Trigger Price   54
  Application of the Model   56
  The Range of Cartel Prices   57
Destruction of the OPEC Cartel: Preconditions and
Possible Causes   59
  Miscalculations of Bargaining Power   60
  Shifts in Country Conditions or Intercountry
  Relations   68
  Shifts in External Market Factors   70
The Long-Run Competitive Price for Petroleum   71
  Production Costs   72
  Transportation Costs   73
  Payments to Resource Owners   73
  The Delivered Price of Oil: The Competitive Case   75
Judgments on the Future World Price   76

4.   TARIFFS, QUOTAS, AND SELF-SUFFICIENCY   79
Economic Effects of Import Controls   80
Tariffs versus Quotas   86
  Price versus Quantity Control   86
  Discrimination   90
  Administrative Flexibility   91
  Scarcity Value of Imports   91
  Conclusion on Tariffs versus Quotas   92

5.   DIRECT POLICIES TO
ENHANCE ENERGY SECURITY   93
Protection Against Supply Interruption   93
  Protection Through Storage   94
  Protection Through Shut-in Capacity   95
  Protection by Combined Storage and Standby
  Capacity   97
  Implementation of Storage and Standby Capacity
  Programs   98
Subsidies for Energy Production   99
  Government Absorption of Energy Production Risk   100
  Payment of Direct Subsidies   104
Ancillary Policies to Increase Energy Security   106
  Federal Land Leasing Policy   106

Deregulation of the Natural Gas Industry   107
Policies to Reduce Energy Consumption   108
  Selective Restraint of Energy Consumption   108
  Energy Consumption Tax   109
Conclusions   110

6.   INTERDEPENDENCE VERSUS
   INDEPENDENCE AS A POLICY OPTION   112
   A Positive Program for Interdependence   115
   Domestic Energy Policy and Interdependence   118
   The Cartel Price with Interdependence   119
   Revised U.S. Import Projections and Their Implications   121

Summary of Review Seminar   126

# Tables

2-1  National Petroleum Council Energy Consumption Projections
    by Sector   16
2-2  Potential Domestic Energy Supply Availability in 1985
    Compared to Actual 1973   17
2-3  National Petroleum Council Projected Import Requirements in
    1985   19
2-4  Sources of U.S. Oil Imports, 1973 and Estimated for
    1985   22
2-5  Estimated U.S. Price–Import Relationship in 1985   26
3-1  Projected Long-Run Cartel Price f.o.b. Each OPEC
    Country   44
3-2  Ordering of Factors Affecting the Internal Discount Rates for
    OPEC Countries and Discount Rates Selected   48
3-3  Present Revenue Equivalent, Cartel Rent, and Lower Bound
    Cartel Price for OPEC Countries   56
3-4  Cartel and World Market Shares and Reserve Shares of OPEC
    Countries, 1970 and 1973   62
3-5  Summary of Bargaining Power Indicators, OPEC
    Countries   65

4-1 Estimated U.S. Price-Import Relationship in 1985　　81
4-2 Opportunity Costs of Import Controls Under Different World Prices and Domestic Conditions　　85
5-1 Storage and Standby Capacity Costs for Energy Security　　98
6-1 Comparison of Present Revenue Equivalent, Cartel Rent, and Lower Bound Cartel Prices for Selected OPEC Countries Under Independence and Interdependence　　122
6-2 Estimated U.S. Oil Production, Consumption, and Imports at a $5.50/bbl Price　　123

# Figures

2-1 Supply and demand for crude oil in 1985　　23
2-2 Using a subsidy to achieve self-sufficiency　　28
2-3 Using a consumption tax to achieve self-sufficiency　　29
3-1 Cartel and competitive pricing strategy　　40
4-1 Effects of import controls　　83
4-2 Comparing a tariff and a quota with an increase in demand　　88

# Preface

This volume is the first printed result of research begun several years ago when RFF made a grant to Paul Homan and Wallace Lovejoy to review the history and rationale of U.S. oil import policy. The death of both authors, in the summer of 1969 and spring of 1970 respectively, robbed the world of two fine economists with a rare feel for the intricacy of the energy world, and brought the project to a standstill. Luckily, my predecessor Sam Schurr was able to persuade Douglas Bohi and Milton Russell to continue the work where Homan and Lovejoy left off.

The traumatic events that followed the October 1973 war in the Middle East suggested that contemplation of the past was, for the time being at least, both less rewarding and less urgent than an analysis of what path the nation might choose in the future. Consequently, a joint decision was made to delay completion of the entire volume and concentrate on dealing with the outlook for oil imports.

In what we are placing before the reader now the focus is on the role that oil imports might play in the foreseeable future (one uses that word with increasing caution, having experienced the scope and sudden impact of the unforeseeable!). Perhaps the core of the analysis is an attempt to put some flesh on the bones of the skeleton called "the future of OPEC." There is abundant speculation why this or that oil-exporting country might or might not be interested in high or low prices, increased or reduced output, close or loose cooperation. Size of reserves, capacity to absorb imports, costs of production, foreign exchange needs, and various other factors—cultural, economic, and social—are cited to support each hypothesis.

Bohi and Russell have attempted to shape assorted characteristics into a composite indicator or predictor of country behavior and have used the sum total of such behavior profiles to hazard a guess at OPEC's cohesiveness and life expectancy. Many persons will quarrel with specific assessments of individual components as well as with the composite outcome. That is as it should be; and RFF, as is its normal position, is underwriting only the quality of the authors' work, not their conclusions.

While the authors have emphasized, in a workshop on the analysis, that at the moment methodology is more important than specific figures or assessments, no one can misread their judgment that the costs of finding accommodation and security in policies involving mutual benefits between exporters and importers will be less than the costs of policies that seek confrontation and all-out independence. The reader will have to judge for himself whether and to what extent the economies of the industrialized countries, and particularly the United States, can attain security along the lines suggested by the authors. They may disagree with the methodology or with the numbers, or with both, or the conclusion may run counter to their intuition; but in any event we trust they will recognize and, one hopes, welcome the attempt to substitute method for musing.

Special thanks are due to the National Science Foundation for funding the one-day workshop at which the thoughts expressed in the manuscript underwent critical review; to Paul Craig, of the staff of the Science Adviser to the President, for his efforts in making the workshop possible, and to the participants in the lively exchange of views.

Washington, D.C.  Hans H. Landsberg
February 1975   Director, Energy and
        Minerals Program
        Resources for the Future, Inc.

# Acknowledgments

Our work on this project started with a research grant from Resources for the Future, Inc. to support the writing of a history and analysis of U.S. oil import policy. Additional support came from a committee chaired by Congressman Thomas M. Rees, the Ad Hoc Subcommittee on Domestic and International Monetary Aspects of Energy and Natural Resource Pricing, of the House Committee on Banking, Currency and Housing. A report was published by this subcommittee under the title, "Oil Imports and Energy Security: An Analysis of the Current Situation and Future Prospects." The study presented here is in part a condensation, refinement, and reorganization of some of the material presented in that report, but the focus of each book is different, as is the audience to which each is addressed.

We gratefully acknowledge the assistance of Nancy McCarthy Snyder, co-author of the subcommittee report. Charles G. Stalon, a colleague at Southern Illinois University, has provided helpful criticism of this work throughout its preparation. We would also like to thank the committee staff, and especially Jane D'Arista, Ben W. Crain, and David Weil, for their assistance.

This manuscript received considerable helpful criticism from the participants in a seminar held in October 1974, co-sponsored by the National Science Foundation and Resources for the Future. A brief summary of some of the comments made there is included at the end of the book. The work was strengthened by changes resulting from the comments of the seminar participants, although some of them would no doubt like to see the end product changed more than it was. The authors, of course, bear full responsibility for the final content of this book. We gratefully acknowledge the support of NSF and RFF which made that seminar possible.

This work was completed prior to the publication of the Federal Energy Administration's *Project Independence Report*. Though the material covered is in part common, the methodology and approach are fundamentally different.

We wish to recognize the patience and understanding of Karrie, Heidi, and Jim and Vicky, Gayla, and Jim while this manuscript was being prepared.

Finally, we are most grateful for the support and encouragement of Hans Landsberg, Director of the Energy and Minerals Program, Resources for the Future.

February 1975                    Douglas R. Bohi
                                 Milton Russell

# 1

# Overview

The development of energy shortages during the past few years, highlighted by the Arab oil embargo, has prompted the United States to consider increased energy self-sufficiency an important domestic policy aim. At one extreme, Project Independence has come to mean, for some persons, ability to forgo completely imports of energy. Others, however, are solely concerned with continuous access to adequate energy supplies. For them, the issue is the best means for achieving energy security.[1]

It is the thesis of this work that energy policy should focus on two goals, energy security and an optimal level of energy consumption. These two goals should be carefully distinguished. The first of these, energy security, would assure a continuous supply of energy to consumers at prices that would not fluctuate with each change in the international political situation. The essence of sovereignty is that access to vital goods not be jeopardized by foreign governments. If the contrary situation exists, then foreign policy must forever be hedged by threats to domestic economic vitality. However, while increased domestic output and reduced domestic energy consumption lead in some sense to increased energy security, these policies

---

[1]This book was written before the *Project Independence Report* (Federal Energy Administration, November 1974) was issued, and the work which went into that document was not available to the authors. "Project Independence" is used throughout to refer to achievement of the announced goal of the Nixon–Ford administrations to reduce imports as a means to energy security. It does not refer to the *Project Independence Report* or to policies outlined therein.

1

are very expensive. Moreover, they could also set in motion forces that may tend to support the price of oil in world markets. Consequently, the energy policy question should take into account the price of energy and the quantity that will be consumed.

The second goal is to reduce, or at least to limit, the transfer of income from oil-importing nations to oil-exporting nations and generally to reduce the cost of energy to consumers. Each of the aims is distinct—energy can be cheap but insecure, or expensive but certain. Both conditions were manifested at the same time beginning in the fall of 1973, but this historical connection should not beguile us into joining the problems analytically. A policy that admirably suits one goal (as self-sufficiency does energy security) may be antithetical to the other (as again self-sufficiency is to lower cost energy).

The drive toward autarchy should be analyzed in much more critical terms than it has been to date. This conclusion follows from two propositions. The first is that the economic damage from politically motivated interruption of supplies can be more efficiently reduced through storage and shut-in capacity than through increased self-sufficiency. The second is that potential monopoly exactions from oil-producing nations can be ameliorated more certainly through a policy that recognizes the desirability of continued oil imports than through a policy of autarchy. The latter proposition rests on the ability to integrate the economies of the separate oil-exporting nations with U.S. capital markets. Moreover, if world oil prices fell from monopoly heights in response to improved international integrating devices, the burden of oil price escalations on the less developed countries would be reduced and a more efficient allocation of the world's resources achieved.

## THE ENERGY SHORTAGE AND PRICE ESCALATION

It is important to note at the outset that the fundamental cause of the current energy shortage was not the Arab embargo of 1973–74. The shortage existed before the embargo, and resulted from several unconnected policy decisions and random events. The price increase for imported oil before the October actions was the result of economic factors which altered the expectations of the world's oil exporters and hence changed their reservation price for oil. Those expectations may be reversed, and if they are, imported oil prices may again fall.

At the time of the oil price surge, the Organization of Oil Exporting Countries (OPEC) had not yet acted as a cartel. It had not limited total production and allocated output among members in order to maintain prices above production costs. It did not have to—world demand for petroleum products had been growing faster than output capacity. Actual OPEC output continued to increase up to the time of the October War, but it did not grow fast enough to keep up with a worldwide consumption surge of 6–8 percent per year. The inadequate growth in production was not due to direct OPEC cartel action. The politically motivated reductions in output and the embargo did, of course, exacerbate the oil shortage when they were implemented, and made the price increases of late 1973 and 1974 possible.

It may be said that OPEC's bark had more effect than its bite. Whereas OPEC's *actions* had not been deleterious, its *pronouncements* created great uncertainty in the international oil market. As a consequence, international petroleum companies shifted their investment away from the low-cost, highly productive, and well-known petroleum provinces in politically insecure countries into areas geologically more risky and more expensive. The new exploration targets were, however, more secure politically and provided diversified potential supply areas. Nevertheless, the major cause of the increase in world oil prices was the pervasive increase in demand resulting from a worldwide economic boom. In the past, business cycles in the important consuming countries tended to alternate so that supplies could be diverted from slumping countries to booming countries. In 1972 and 1973 simultaneous boom conditions developed.

The United States in particular placed large extra demands on world petroleum capacity. Crude oil and product imports averaged 6.1 million barrels per day (MMb/day) in 1973 compared to 4.7 MMb/day in 1972 and 3.9 MMb/day in 1971. The monthly averages throughout 1973 and 1974 consistently exceeded those of 1971 and before, despite the embargo. Why did U.S. demand for imported oil increase so dramatically? A facile answer is that more stringent environmental regulations led to increased demand for "cleaner" fuels, such as petroleum, especially in the electric generation and transportation sectors. Evidence suggests, however, that a combination of other factors was more significant. These factors include the delayed effects of state oil prorationing controls, the impact of the oil import quota in restraining refinery growth and hindering energy planning, price controls, natural gas price regulation, and varied policies regarding coal and nuclear power.

Prorationing and the import quota created a stable price for crude oil for the two decades prior to 1971. Prorationing diverted investment toward development drilling and away from exploration in more risky but potentially more productive provinces. The excess capacity associated with production controls was not based upon a deepened reserve base that could sustain production when called upon. Consequently, as oil demand in the United States increased in the late 1960s and early 1970s, there was inadequate inducement to take the necessary steps to increase output. The price explosion that followed had not yet had time to be translated into either long-run demand or supply responses. Timely price signals had been dampened by prorationing and import controls; thus the adjustment process was hindered.

Import controls also discouraged necessary investment in domestic refining capacity before 1973. Refiners could not rely on domestic production as a source of feedstock, nor on foreign crude because of import limitations. Adequate domestic refining capacity was not available when the import controls were lifted and U.S. refiners had access to foreign crude supplies. Refining capacity that was made available by declining U.S. output often was not suitable for the high-sulfur foreign crudes available.

Additional inflexibilities existed in related areas. The increase in world oil demand placed new burdens on tanker capacity. The steel industry had little advance warning of the surge in demand for tubular goods and drilling equipment. That demand was superimposed on a steel industry already operating at capacity due to the capital goods boom of 1972–73. Price controls removed some of the incentive to respond to the needs of industries supplying energy and interfered with the efficient allocation of the available steel among potential users. Price controls prevented the natural gas market from reflecting the deteriorating gas reserve position, and hence neither supply nor demand patterns reflected appropriate allocation of this resource. Reverberations were wide—consumers were drawn to gas, away from coal, and the problem of making coal an environmentally sound fuel was ignored. When the gas *delivery* situation suddenly deteriorated (after a long decline in the quality of reserve coverage), it exacerbated the oil and liquid petroleum gas (LPG) shortfalls.

The coal and nuclear industries entered this period of energy

stringency unequipped for large increases in output. The coal industry was faced with heightened environmental standards, an inadequate transportation system, shortages of essential raw materials
and skilled manpower, and increasingly stringent health and safety
standards. Uncertainties as to future markets hampered the industry in moving forcefully to meet these conditions with timely action.
In nuclear energy, which is faced with some of the same uncertainties, there have been delays far longer than anticipated in bringing
plants on-stream. Viewed against the background of the recent
energy situation, the meager results of the nuclear program appear
worse, perhaps, than they really are. Clearly, though, the nuclear
energy sector has not fulfilled the promises made for it. The U.S.
commitment to nuclear power also hindered the development of alternative energy sources; it especially subverted the drive for better
means to mine and use coal. In view of the problems that have
arisen in the 1970s, the nuclear program, with all its ramifications,
has been a near disaster.

The convergence of these circumstances led to unprecedented
energy price escalations. Constraints imposed by government and
by the private sector perhaps prevented even further price increases. As a consequence, the market for energy did not clear,
shortages developed, and nonprice allocations took place. It was because of these price increases and shortages, but perhaps in ignorance of their cause, that concern about energy security arose. The
drive for self-sufficiency in turn was fueled in the popular mind by
the impression that our energy problems arose primarily because of
exporting country actions.

### Routes to Energy Security

This work compares two approaches to the issue of energy security. The first of these is the familiar one that energy security is
equated with reduced reliance on foreign energy supplies. The second derives from the view that economic interdependence can bring
long-run energy security and lower energy costs. The latter argument is developed primarily in the final chapter, where policies are
suggested which would lead to differences in expectations, and thus
in behavior, of both the United States and the oil-exporting nations.
The analytical framework and factual basis common to both approaches is developed in chapters 2–5. These chapters represent an
analysis of the policy options available to the United States so long

as self-sufficiency remains the touchstone of U.S. energy policy. Chapter 6 takes a broader view of the U.S. interest; using the analysis developed early in the work, it outlines the effects that would result from the efforts of the United States to integrate its energy economy with that of the oil-exporting countries.

The study begins with a set of energy supply and demand projections for the United States in 1985.[2] These projections are based on studies by the National Petroleum Council, with certain modifications. Any gap between domestic supply and demand at a given price implies either imports or an increase in the price. The base level of import requirements is estimated assuming a domestic price of $7.88/bbl (1973 prices). The import projections are broken down by expected country of origin. Following the "security through autarchy" approach, estimates are then derived of the increase in the domestic price that is necessary in order to restrict the level of imports. It may be convenient to think of these price increases as resulting from either a tariff or a quota. Similar import restrictions may be achieved by production subsidies or by a consumption tax.

The price–import relationship cannot be translated into policy options without assumptions as to future world oil prices. To this end, chapter 3 analyzes the forces operating to determine the future course of the OPEC cartel.[3] The analysis produces a plausible range of future world prices, based on the assumption that the United States will follow a positive policy to restrict reliance on imports. That range is bounded on the lower end by a competitive market price and on the upper end by the self-sufficiency price in the United States. Within this range, it is argued, a lower bound cartel price can be identified. The long-run world price will tend toward this lower bound if the OPEC cartel remains effective indefinitely.

Policy options available to the United States to achieve energy

---

[2]The forecast date is 1985 rather than the initially announced Project Independence target date of 1980. This adjustment was made primarily because of increasing agreement that the cost of achieving self-sufficiency by 1980 would be intolerable. The lags in energy supply increases imply that self-sufficiency could be achieved by then only through so dramatic a reduction in energy consumption as to disrupt the American life-style.

[3]The term "cartel" is used as a shorthand reference to OPEC price-influencing behavior even though the organization does not fit the restrictive definition of a cartel per se. Alternative descriptive terms are cumbersome. Moreover, the use of the word "cartel" in this connection is so widespread in the press and in other writings that attempts to be more precise may tend more to obfuscate than to clarify. Neither legal nor pejorative inferences should be drawn from the use of this word.

security are reviewed in the context of projections of U.S. output and consumption and expected world oil prices. First, however, the nature of the threat to security must be examined. The national security issue regarding oil imports is often divided into political and economic aspects. The political threat is said to arise from the leverage that suppliers of foreign oil can exert on U.S. foreign policy. Three points should be made in this regard. First, political leverage exists only because of potential economic harm. The political vulnerability of U.S. imports is dominated by economic vulnerability. There is no need to distinguish between the two. Second, the interruption of oil supplies is a problem fundamentally distinct from the issue of cartel monopoly exactions. Interruption is (a) a political act which, by its nature, cannot be expected to gain the support of all oil-exporting nations; (b) a short-term action, the effectiveness of which begins to fade immediately upon its inception; and (c) a threat that can be countered in a multitude of ways. Finally, potential harm to a friendly state could create sufficient pressure on the United States to cause it to alter its foreign policy. This troublesome feature of the energy security problem implies that if friendly nations are unwilling or unable to undertake the steps necessary to reduce their jeopardy, the United States must accept that burden.

The economic aspect of energy security, as usually discussed, involves monopoly exactions by foreign producers, price uncertainty, and an unfavorable balance of payments. Self-sufficiency ultimately produces the same long-term effects as monopoly exactions by foreign producers. It forces domestic prices high enough to increase domestic production and reduce consumption, so that the market clears. An autarchic policy enables a nation to follow a more efficient path to the same end than would be possible if imports were suddenly interrupted. Consequently, the problem of monopoly exactions cannot easily be dealt with in the energy security context. The second problem, uncertainty about world prices, leads to reduced willingness of U.S. industry to develop alternative energy sources, and hence reduces energy security. The solution becomes one of providing U.S. producers with partial protection against sharply declining prices as one element of energy security policy. The balance-of-payments issue is more complex and requires greater attention, though it is concluded here that it need not be particularly troublesome for the United States except indirectly as other countries experience foreign exchange problems.

The United States must finance oil imports, and the rising price of petroleum has caused much concern about the implications for the balance of payments if imports are uncontrolled.[4] If the United States does not limit oil imports and the oil-exporting nations respend their dollar earnings in the United States, there is no foreign exchange burden. A problem with foreign exchange develops when the oil-exporting nations earn more dollars than they want to respend in the United States. There are three alternatives for disposing of their dollar earnings: they can trade them for other currencies or gold; they can invest the dollars in the United States; or they can simply hold the dollars as an asset.

If dollars are exchanged for other currencies (in order to purchase goods from, or invest in, another country) or for gold, the dollar could be weakened in world markets and the U.S. terms of trade with the rest of the world worsened. The amount of U.S. goods that must be exchanged to obtain a given amount of imports would rise, increasing the amount of U.S. resources needed to support all imports in addition to oil. It is this expectation that, for balance-of-payment reasons, lends support to the self-sufficiency approach to energy security.

If dollars were invested in the United States, the oil-exporting nations would be buying U.S. assets rather than U.S. goods. The effect of purchasing assets is the same as purchasing goods as far as the foreign exchange market is concerned. No balance-of-payments problems exist. Moreover, if the oil-exporting nations can be induced to exchange their natural assets for American capital assets, a potent constraint against the imposition of an oil embargo is introduced. In addition, safe outlets for capital investment provide an incentive for key oil-exporting nations to increase oil sales by reducing the reservation price at which they will transfer oil. This relationship is explored in chapter 6, where the concept of interdependence is formulated.

The final option for disposition of foreign exchange earnings by oil-exporting countries is for them to hold their dollar earnings as a liquid asset. Such behavior has no immediate effect on foreign ex-

---

[4]A self-sufficiency policy potentially involves the same kind of problem. If high prices were imposed on the United States while the rest of the world enjoyed low energy prices, U.S. energy-intensive exports would be less competitive abroad and foreign goods more attractive in the United States. On the other hand, if the rest of the world paid monopoly prices and the United States achieved self-sufficiency, the dollar would be stronger relative to the currencies of the oil-importing countries, creating other strains on international monetary mechanisms.

change markets, but it does pose a potential danger to the entire monetary system. A large accumulation of dollars (and other currencies) in the hands of a small group of countries could be used for political or economic blackmail. The danger to the monetary system could be minimized, however, by cooperation among the major financial centers. If, for example, the oil-exporting nations decided to dump their dollar holdings in international money markets in exchange for Deutschmarks and yen, the potential disruptive effect would be neutralized if German and Japanese monetary authorities chose to accommodate the exchange and add the dollars to their reserves. No exchange rates need be changed; the oil-exporting countries would have simply exchanged their dollars for another currency. If the oil exporters then decided to sell their mark and yen holdings, Germany and Japan could use their newly acquired dollars to purchase them back. Otherwise, an accommodation among the United States, Germany, and Japan could be reached with regard to the outstanding dollars. These transactions require the active participation and cooperation of official central banks; they cannot be handled by private institutions alone.

The preceding analysis is based on the expectation that a balance-of-payments deficit on current account will be created with regard to oil exporters. A more serious problem arises if the oil exporters conclude that it is not in their interest to sell oil in an amount that creates balance-of-payments surpluses. Such a decision would drive up the price of oil and leave consumers worse off. The failure to recognize that instability in world capital markets may also create a balance-of-payments "problem" for surplus countries has led to sometimes myopic views of the world economy.

### SUMMARY OF RESULTS

The nature of the energy security problem is outlined above, but its dimensions, and an evaluation of the policy options to meet it, can only be understood when projected consumption, output, imports, and price are considered. It is the finding of this study that if the domestic price of crude oil persists indefinitely at $7.88/bbl, and no new policy initiatives are adopted, a midpoint estimate of 1985 oil imports is about 8 MMb/day (or 35 percent of consumption). Under more pessimistic supply and demand conditions, imports may run as high as 15 MMb/day (or 55 percent of consumption) at the upper extreme. This range of import risk is taken into account by the security policy options considered in the self-sufficiency framework. The Arab countries are the import sources having the highest risk. The Arab share of the two import estimates would be roughly 2 MMb/day and 8 MMb/day respectively. The lower estimate duplicates 1973 experience (about 8 percent of consumption), while the upper estimate implies a major increase in dependence on Arab oil (to about 30 percent of consumption).

The dependence of the United States on imports would be reduced if the domestic price of crude oil were higher than $7.88/bbl. Depending on the assumptions used, and again without the enactment of other policies, it is estimated that imports could be reduced to about 20 percent of domestic production (about 17 percent of consumption) if the price were expected to remain at $9/bbl in the lower case, or about $11.30/bbl in the upper case. Complete self-sufficiency is estimated at domestic prices ranging between $10 and $13/bbl.

These price estimates imply that U.S. self-sufficiency can be achieved at about the current landed price of foreign oil. Current world prices would have to be expected to prevail indefinitely, of course, to permit the necessary long-run adjustments. The analysis in chapter 3 suggests instead that prices may be expected to fall. How far and how fast are crucial questions. If the world market became highly competitive, the landed U.S. price could fall to as low as $3/bbl. This result is not likely. On the basis of the analysis below, if current policy is followed, the OPEC cartel is not expected to collapse at least before 1985, and the landed price of foreign oil will fluctuate above $7.50/bbl.

The $7.50 lower bound cartel price is about the same as the assumed price used in the original import projections. This implies that the base case import projections are consistent with the absence of import controls. A $3/bbl import price implies highly restrictive import controls (an import duty of nearly $5/bbl), even in the base case. The opportunity cost of import controls to U.S. consumers and to the economy depends on the gap created between domestic and import prices. If all imports were eliminated by 1985, the opportunity cost to U.S. consumers might reach $66 billion per year if the world price fell to $3/bbl; on the same assumptions, it would reach $34 billion per year if the world price were $7.50/bbl.

The conclusion from this analysis is that complete self-sufficiency is an unacceptably costly target unless the world price remains about $10/bbl. It is also unnecessary, even within an autarchic framework, if one considers that the bulk of future U.S. imports may come from relatively secure sources. There would be no need to restrict imports from insecure sources either, if sufficient storage or standby reserves were maintained to ensure against the harmful effects of short-term interruption.[5]

In chapter 5 cost estimates for appropriate levels of storage and shut-in capacity are developed. Storage of a 6-month supply (in steel tanks with pipeline hookups) of projected insecure (Arab) imports in 1985 is estimated to cost approximately $0.7 billion per year at the low estimate of imports and $2.9 billion at the upper estimate. These estimates are based on a price of $7.50/bbl of oil, with the tankage investment costs amortized over 15 years at a 10 percent interest rate. If the world price should fall below $7.50, increased storage would be required to cover the influx of Arab imports. Part of the costs would be offset by the cheaper price of oil.

The annual cost of standby capacity to cover projected Arab imports would be approximately $3.0 billion at the low estimate of imports and $13.1 billion at the higher level. While more expensive than storage, standby capacity is attractive when compared with the costs of self-sufficiency. Even more attractive is a policy combining storage of 3 months of imports and substantial standby capacity. The annual cost of a combination policy which would provide appropriate security against interruptions would be about $1.2 billion at a low level of imports and about $5.7 billion if imports were at their upper limit. The combination policy also has benefits beyond protecting against use of oil for political ends (see chapter 5).

A variety of subsidies and taxes may also be used to stimulate domestic production or reduce domestic consumption. General subsidies to private industry are not recommended because of the inefficient resource allocation that results. Selective support of research and development, improvement in the functioning of energy markets, and changes in federal leasing policy are found to be desirable actions that efficiently reduce reliance on foreign energy sources. An energy consumption tax is recommended, but

---

[5]"Standby reserves" and "shut-in capacity" describe the ability to produce crude oil from existing facilities. This capacity can exist either because reservoirs are produced at less than capacity rates or because reservoirs are ready to be produced but are held out of production. In either case, transportation and other facilities must be in place to handle the larger production if the extra capacity is to be included as part of the emergency reserve.

only at a level necessary to cover the cost of government-supported, energy-related programs. A standby tariff system is the preferred means for limiting the risk to the domestic industry of a major decline in world prices.

In summary, a combination of programs is required if energy security is to be achieved efficiently through reduced reliance on foreign energy sources. The political threat of embargo can be negated by a program of enhanced oil storage. Shut-in capacity would provide the bridge between the short-run response of drawing down oil stocks and the longer-run adjustment of domestic and worldwide supply and demand patterns to changed availability of imported oil. Overall reliance on foreign energy could be reduced by encouraging domestic energy production through government-supported research and development and through standby tariffs. An energy consumption tax to finance the necessary government ventures and to reduce energy consumption would also be appropriate. Risk could thus be reduced to an acceptable level without driving the nation to autarchy, even if security were identified with decreased reliance on imports.

A view of the world which contrasts with the autarchic posture is posited in chapter 6. There the concept of energy and economic interdependence between the United States and the oil-exporting countries is explored. With interdependence, the investment alternatives (and thus the perceptions of the oil-exporting and consuming countries) would be significantly revised, with far-reaching results for the world oil market. New calculations based on the analysis in chapter 3, but with different assumptions about the perceptions of the OPEC countries, yield a revised range of potential cartel prices. The upper limit remains at approximately $10, but the lower limit is reduced to near $5. Again, of course, the apparent precision of these estimates should not mislead readers; the relative magnitudes involved are the important aspects of the study. Interdependence as described in chapter 6 would not alter the competitive price for oil nor the possibility for breakdown in cartel control. The result would be that interdependence, while leading to a *lower* possible world cartel price for oil, would not necessarily lead to a more insecure cartel. This follows from the view that the cartel can adjust to marginal shifts in U.S. demand. Interdependence would slow the development of conventional fuel replacement technology. It would thus reduce the risk to the cartel of sudden market dislocation.

The implications of these changes for U.S. consumption, production, imports, and prices are further explored in chapter 6. The general outlines are obvious. Production in the United States would be lower, consumption higher, and imports would supply a larger proportion of the total energy used. Nations without the self-sufficiency option most likely also would find their energy costs lower. Unfortunately, the increased reliance on foreign energy would carry with it an increased need for protection from politically motivated supply interruptions, with consequent increased expense. It is the essence of interdependence that the balance-of-payments problems would be reduced because of the more ready recycling of petroleum revenues.

Before the interdependence option can be explored, the basic structure of the energy market must be presented. This task, and an interpretation of energy security options in a self-sufficiency framework, are taken up in chapters 2–5.

# 2

# Future Import Requirements and Domestic Prices

It is necessary to estimate the magnitude of future imports and the countries from which these imports would be obtained in order to assess the security risk associated with importing oil. The objective of Project Independence is to reduce the potential threat to national security arising from a dependence on imported energy supplies, but independence can be defined in terms of an acceptable level of imports, in terms of acceptable sources of imports, or in terms of the capability to substitute domestic supplies or standby reserves in the event of interruption of a foreign supply. However defined, the prospective volumes and sources of imports are directly related to the security issue, and must be estimated.

The estimates here are based on work already published by the National Petroleum Council (NPC).[1] A desirable feature of the NPC projections is that a range of possible outcomes is presented, which depends on alternative assumptions concerning the future. Furthermore, the projections are presented in sufficient detail, with supporting material, that the assumptions can be altered in order to modify the final estimates.

The approach here is to assume that a given price of crude oil will prevail between now and 1985, and then to project the amount of domestic production and consumption that will take place at that price. The assumed price is $7.88 per barrel (in 1973 prices); this price level corresponds to the average of the 1985 price assumptions

---

[1]National Petroleum Council, *U.S. Energy Outlook* (December 1972).

14

in the NPC forecast for what is termed supply Cases I and III, adjusted to 1973 prices.[2] It also corresponds roughly to the average price of domestic oil in the United States at the time of this writing (an average of $5.25/bbl on "old" oil and about $10/bbl on "new" oil as of August 1974). The forecast, therefore, supposes that the current average price will prevail into the future and attempts to determine how much output and consumption may be expected in 1985 at this price.[3]

It is assumed further that imports will make up the gap between domestic supply and demand at the assumed price. This implies either that the world price will be around $7.88/bbl or that the world price will be below $7.88/bbl, but imports will be limited to the gap between domestic output and consumption at the $7.88 price. This restriction will be dropped in order to determine the amount by which an increase in the domestic price will reduce imports because it increases output and reduces consumption. This approach avoids the difficulty of simultaneously projecting domestic supply, demand, and the corresponding equilibrium price. It is satisfactory as long as imports can be obtained to close the gap without necessitating a change in the domestic price.

### U.S. SUPPLY–DEMAND BALANCE: THE BASE CASE

The NPC consumption projections for 1985 are given in table 2-1 for three rates of growth: low, intermediate, and high. All three estimates are based on the assumption that the downward trend in energy consumption per dollar of gross national product (GNP) and the upward trend in energy consumption per capita will continue into the future. In addition, the low growth rate assumes a relaxation of environmental standards in the future, adoption of technological improvements in energy efficiency at a faster rate than in the past two decades, more concentrated settlement patterns, and

---

[2]This price could be rounded to, say, $8/bbl. It was decided against rounding this and subsequent forecast figures in order to enable the reader to duplicate the results.

[3]In order that the $7.88 price may prevail in the long run, the marginal and average prices must be the same. The $10 price is not a true marginal price because it cannot be earned on marginal output from all wells and because of the adverse effects of price controls on production incentives. Thus, the true marginal price is somewhat less than $10, so the assumption of equality between the average and the margin may not be excessively heroic.

TABLE 2–1.   National Petroleum Council
Energy Consumption Projections by Sector

| | Residential and commercial | Indus-trial | Transpor-tation | Electrical conversion | Other | Total |
|---|---|---|---|---|---|---|
| A. Annual growth rates | | | | | | |
| 1960–1970 Actual | 4.0 | 3.4 | 4.2 | 7.2 | 3.4 | 4.3 |
| 1970–1985 Projected | | | | | | |
| Low | 2.8 | 2.1 | 3.3 | 5.7 | 4.7 | 3.4 |
| Intermediate | 3.5 | 2.9 | 3.7 | 6.6 | 5.3 | 4.1 |
| High | 4.0 | 3.1 | 3.9 | 6.9 | 5.5 | 4.4 |
| B. Annual volume (quadrillion Btu)[a] | | | | | | |
| 1970 Actual | 15.8 | 20.0 | 16.3 | 11.6 | 4.1 | 67.8 |
| 1985 Projected | | | | | | |
| Low | 23.9 | 27.1 | 26.7 | 26.7 | 8.1 | 112.5 |
| Intermediate | 26.6 | 30.9 | 28.3 | 30.2 | 8.9 | 124.9 |
| High | 28.5 | 31.9 | 29.0 | 31.4 | 9.2 | 130.0 |

*Source:* National Petroleum Council, *U.S. Energy Outlook*, p. 37.
[a]All numerical units in this work are American, i.e., a quadrillion is $10^{15}$.

improved transportation systems. The high demand case, on the other hand, assumes more stringent environmental standards than under consideration as of 1972, slower development and implementation of technological improvements, a continuation of past urbanization trends, and the same life–style that has developed in recent years.

Broken down by energy-consuming sectors, residential–commercial consumption is not expected to grow any faster than the average annual rate of the past 10–15 years, even with the highest projection. A slowdown may be expected on the basis of trends toward smaller dwelling units and more apartments because of the rising proportion of young adults, fewer children per family, and rising construction costs.[4] The range of projections for industrial energy use is well below actual growth rates in recent years. The projected reduction is based on expected increases in energy efficiency. The transportation sector is also expected to grow less rapidly in the future, from a 4.2 percent rate during 1960–70 to between 3.3 and 3.9 percent to 1985.[5] This change is based on a shifting demographic structure, a larger proportion of economy cars, and a resurgence of

[4]*U.S. Energy Outlook*, pp. 41–44.
[5]*U.S. Energy Outlook*, pp. 46–48.

railroads and other mass transportation systems. Electrical conversion refers to the energy loss that occurs in converting the British thermal unit (Btu) content of fuels into electrical power. The forecast anticipates an increase in the efficiency of electrical conversion at the generation and transmission stages.

It should be noted that conditions have changed considerably during the 2 years since the NPC forecast was developed. In particular, the forecast assumed energy prices in 1985 that are already in existence today. The NPC forecast anticipated that these higher price levels would be reached gradually over the next decade. In every sector these higher energy costs may be expected to reduce energy consumption growth rates even more than the NPC study projected. The change in price suggests that the low and intermediate consumption forecasts bracket the most likely range of outcomes. Consequently, for the energy supply–demand balance developed here, the high consumption growth rate is discarded.

The projections of potential domestic energy supply are given in table 2-2 for two NPC supply cases (out of several offered in the NPC study). Case I assumes a "high" oil and gas drilling rate and a "high" finding rate, while Case III assumes a "medium" drilling rate and a "low" finding rate. These two cases appear to bracket the likely outcomes that are consistent with the urgency of developing domestic oil and gas supplies. The high drilling rate is the only assumption consistent with a national push toward self-sufficiency, because this

TABLE 2-2. Potential Domestic Energy Supply Availability in 1985 Compared to Actual 1973

|  | 1973 Actual | 1985 NPC Projections | |
|  |  | Case I | Case III |
| --- | --- | --- | --- |
| Oil, total (MMb/day) | 10.9 | 16.89 | 12.31 |
| Shale | 0.0 | 0.75 | 0.40 |
| Coal conversion | 0.0 | 0.68 | 0.08 |
| Gas, total (Tcf/year) | 22.6 | 34.4 | 22.5 |
| Nuclear | 0.0 | 1.3 | 0.8 |
| Coal gasification | 0.0 | 2.5 | 1.3 |
| Hydroelectric (bil kWh/year) | 280 | 316 | 316 |
| Geothermal (MWe)[a] | 400 | 19,000 | 7,000 |
| Coal (MMt/year) | 589 | 1,093 | 863 |
| Nuclear (MWe) | 80,000 | 450,000 | 300,000 |
| Total (quadrillion Btu) | 68.7 | 131.5 | 93.8 |

Source: National Petroleum Council, *U.S. Energy Outlook*, p. 18.
[a]MWe = Megawatt electrical.

aspect of supply is under policy control. Case III may be regarded as a possible outcome in the event that the most appropriate policies are not undertaken, or if success rates are low. The amounts of geothermal and hydroelectric energy are of relatively minor importance in all cases, while the variance in coal and nuclear energy among the supply cases becomes irrelevant when it is noted that total possible output cannot be completely utilized.

The medium drilling rate for crude oil (footage drilled annually) in Case III assumes that footage reached in 1985 will match the level maintained during 1960–65 (approximately 130 million feet). The low finding rate (volume discovered per unit of drilling) in Case III assumes that oil discoveries will average 100 barrels per exploratory foot drilled, about the same rate achieved during the 1965–70 period, but 10 barrels per foot less than the average from 1950 through 1965. The high drilling rate for crude oil assumes a growth rate high enough to increase footage from the level reached in 1970 (90 million feet) to a level in 1985 that exceeds the industry all-time high achieved in 1956 (nearly 200 million feet). The high finding rate assumes a success rate 50 percent greater than the low finding rate.

For natural gas, a medium drilling rate corresponds to a 3.0 percent annual increase, from 40 million feet per year in 1970 to 65 million feet in 1985. A high drilling rate corresponds to a 5.4 percent annual growth to nearly 90 million feet in 1985. The low finding rate simply continues the actual downtrend of recent years, to 200 thousand cubic feet (Mcf) per foot drilled in 1985, while a high finding rate is uniformly about 50 percent higher than the low rate.

These parameters are applied to oil and gas production in the lower forty-eight states and do not include production from northern Alaska and its offshore waters. It is estimated that about 30 percent of the remaining domestic oil and gas reserves are located in these provinces; 119 billion barrels of oil-in-place and 327 trillion cubic feet (Tcf) of recoverable gas.[6] The basic limitation on potential supplies from Alaska is the delivery system. Both Cases I and III assume that oil production will begin in 1976 and gas production in 1978. In Case I it is estimated that oil production will start at a rate of 750 Mb/day and reach 2.6 MMb/day in 1985; gas production

---

[6]*U.S. Energy Outlook*, p. 95.

TABLE 2-3. National Petroleum Council
Projected Import Requirements in 1985

| | Case I | Case III |
|---|---|---|
| Total potential supply (quadrillion Btu) | 131.5 | 93.8 |
| Unutilized coal and nuclear power (quadrillion Btu) | 20.0 | 3.8 |
| Net domestic supply (quadrillion Btu) | 111.5 | 90.0 |
| Total consumption (quadrillion Btu) | | |
|   Low demand | 112.5 | 112.5 |
|   Intermediate demand | 124.9 | 124.9 |
| Import requirements | | |
|   Low demand (quadrillion Btu) | 1.0 | 22.5 |
|     Oil: quadrillion Btu | 0.0 | 16.1 |
|       MMb/day | 0.0 | 7.6 |
|     Gas: quadrillion Btu | 1.0 | 6.4 |
|       Tcf/year | 1.0 | 6.4 |
|   Intermediate demand (quadrillion Btu) | 13.4 | 34.9 |
|     Oil: quadrillion Btu | 7.5 | 28.5 |
|       MMb/day | 3.6 | 13.5 |
|     Gas: quadrillion Btu | 5.9 | 6.4 |
|       Tcf/year | 5.9 | 6.4 |

Source: National Petroleum Council, *U.S. Energy Outlook*, pp. 30, 32.

(north of the Brooks Range) is estimated to begin at 0.8 Tcf/year in 1978 and reach 3.3 Tcf/year in 1985.[7] In Case III it is estimated that oil production will begin at 600 Mb/day and increase to 2.0 MMb/day in 1985; gas production will start at 0.6 Tcf/year and increase to 2.2 Tcf/year in 1985.[8]

In balancing total supply available with projected demand, it is assumed that all oil, gas, hydroelectric, and geothermal energy that becomes available would be utilized, but not all possible coal and nuclear power. In the case of oil and gas, this assumption is based on the close substitutability of other fuels in all consuming sectors. Hydroelectric and geothermal energy are used only in electric power generation, but available supplies are small. The potential growth in supply of coal and nuclear power is large, but can be absorbed for the most part only in the electric utility sector. The NPC projections assume that up to 74 percent of total electric utility fuel consump-

[7] *U.S. Energy Outlook*, p. 112.
[8] *Ibid.*
[9] *U.S. Energy Outlook*, p. 20.

tion in 1985 will be coal and nuclear (compared to 48 percent in 1970).[9] Even so, all of the potential coal and nuclear supply that could be available would not be utilized.

In table 2-3, the total potential supply is reduced by the estimated amounts of unutilized coal and nuclear energy. The net domestic supply is balanced against the projected levels of low and intermediate consumption to obtain import requirements. The oil import requirement in 1985 is estimated to range between zero (low demand, high domestic supply) and 13.5 MMb/day (intermediate demand, low supply). The natural gas import requirement in 1985 ranges from 1.0 Tcf/year to 6.4 Tcf/year.

The NPC projection for natural gas imports is puzzling. Presumably the forecast assumes that U.S. natural gas prices would be permitted to vary freely so that a 15 percent return on investment is assured at the projected rate of output. This produces an NPC estimate of the field price of natural gas in a range between $0.52 and $0.63/Mcf. Clearly, the domestic price can rise a great deal higher before it becomes economic to import liquefied natural gas. But the NPC forecast also projects maximum pipeline imports from Canada (and Mexico) in 1985 of 2.7 Tcf/year.[10] Thus, any amount over 2.7 Tcf/year in table 2-3 represents liquid natural gas (LNG) imports. The assumption of a free market price of natural gas in the United States seems reasonable (i.e., prices will become deregulated before 1985). It also seems reasonable to assume that the domestic market can be satisfied at gas prices below the landed cost of LNG.[11] Thus, gas imports are not expected to exceed the 2.7 Tcf/year maximum pipeline imports.

For convenience, the gas import requirement will be converted to oil equivalents and added to total crude oil import demand. This brings the estimated range of oil imports to 0.5 MMb/day in the lowest case and 14.8 MMb/day in the highest import case. If the conditions generating the lower and upper limit projections are regarded as equally likely, it follows that the midpoint of both extremes may be regarded as the best point estimate of future import requirements. This amounts to 7.6 MMb/day in 1985.

---

[10]U.S. Energy Outlook, p. 133.

[11]This is consistent with available free market price forecasts. See, for example, Paul W. MacAvoy and R. S. Pindyck, "Alternative Regulatory Policies for Dealing with the Natural Gas Shortage," The Bell Journal of Economics and Management Science, vol. 4, no. 2, Autumn 1973.

## Sources of Projected Imports

In the analysis that follows, emphasis will be placed on two projected levels of import requirements: the 7.6 MMb/day midpoint estimate and the 14.8 MMb/day upper bound estimate. While the range below 7.6 MMb/day may be considered as likely as the upper range, the approach in this work is to emphasize the potential security risk of future imports. Energy security policy must guard against the pessimistic possibilities, and thus the low-import situation is ignored in this analysis.

Table 2-4 gives the expected sources of projected U.S. oil imports and a comparison with the 1973 pattern. The projected pattern is based on a number of assumptions. First, a significant increase in imports from Canada is not expected until production from the Mackenzie Delta and Arctic islands begins; such production is not anticipated before 1985. Some increases in output in South America are expected, but so are increases in local consumption requirements. Thus, additional imports are not anticipated from this area either. Imports from Iran, Nigeria, and Indonesia are expected to increase in the future. These increases are not intended to reflect a corresponding rate of expansion in production capacity, but rather an anticipated policy of favoring dispersion of supply sources away from the Arab countries.

It is concluded, therefore, that any major increase in U.S. imports must come from the Arab countries. If it is assumed further that Libya and Kuwait will continue their policy of restraining the growth of output, the bulk of additional imports must come from Saudi Arabia. As pointed out earlier, the projections assume that imports will satisfy any domestic supply–demand gap. Saudi Arabia may not be as accommodating as the forecast assumes. Note that the pattern of imports for the midpoint estimate closely resembles that of 1973 and the volume remains at about 35 percent of consumption. In the upper estimate, however, imports rise to 54.6 percent of consumption, and the Arab share rises to 29.1 percent of U.S. consumption.

## U.S. Imports at Higher Prices

Given the magnitude and pattern of imports derived in the table, consider next the extent to which these imports may be reduced at

TABLE 2-4.  Sources of U.S. Oil Imports,
1973 and Estimated for 1985

| Source | 1973 Estimate[a] | | 1985 Midpoint estimate | | 1985 Upper estimate | |
|---|---|---|---|---|---|---|
| | Amount (MMb/day) | Percent of consumption | Amount (MMb/day) | Percent of consumption[b] | Amount (MMb/day) | Percent of consumption[c] |
| Venezuela | 2.0 | 11.5 | 2.0 | 9.0 | 2.0 | 7.4 |
| Canada and other W. Hemisphere nations | 1.5 | 8.7 | 1.5 | 6.8 | 1.5 | 5.5 |
| Subtotal | 3.5 | 20.2 | 3.5 | 15.8 | 3.5 | 12.9 |
| Iran | 0.4 | 2.3 | 1.0 | 4.5 | 1.5 | 5.5 |
| Nigeria and other African nations | 0.6 | 3.5 | 1.0 | 4.5 | 1.5 | 5.5 |
| Indonesia and other S.E. Asia nations | 0.2 | 1.2 | 0.3 | 1.3 | 0.4 | 1.5 |
| Subtotal, non-Arab nations | 4.7 | 27.2 | 5.8 | 26.1 | 6.9 | 25.5 |
| Saudi Arabia | 0.5 | 2.9 | 0.9 | 4.1 | 6.0 | 22.1 |
| Libya | 0.3 | 1.7 | 0.3 | 1.3 | 0.3 | 1.1 |
| Kuwait | 0.1 | 0.6 | 0.1 | 0.5 | 0.1 | 0.4 |
| Other Arab nations | 0.5 | 2.9 | 0.5 | 2.2 | 1.5 | 5.5 |
| Subtotal Arab nations | 1.4 | 8.1 | 1.8 | 8.1 | 7.9 | 29.1 |
| Total | 6.1 | 35.3 | 7.6 | 34.2 | 14.8 | 54.6 |

[a]Estimated by adding to published reports of crude oil imports (i.e., Federal Energy Administration) the original source of crude oil for refined product imports. Original source is estimated from crude oil imports of countries where the United States obtains refined products.

[b]As a percent of the average of the low and intermediate consumption projection (22.2 MMb/day).

[c]As a percent of the intermediate oil consumption projection (27.1 MMb/day).

higher levels of U.S. prices. Alternatively viewed, the main concern is the level to which the U.S. price must rise if restrictions are placed on oil imports.

The situation in 1985 is depicted in figure 2-1. The functions are intended to reflect the long-run supply and demand for crude oil in the United States at that time, as set out in the preceding section. At the $7.88/bbl assumed price, domestic supply is given as $Q_1$, domestic demand is $Q_2$, and the gap between them represents the demand for oil imports. The amounts represented by $Q_1$ and $Q_2$ correspond to the alternative estimates derived above. If the price were to rise above $7.88/bbl, the quantity supplied would increase above $Q_1$ as

FIGURE 2-1.    Supply and demand for crude oil in 1985.

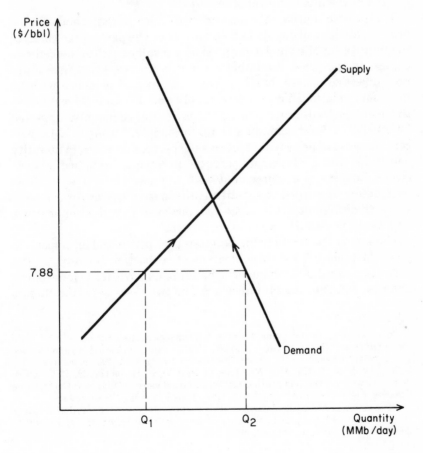

we move up along the supply curve. The quantity demanded would at the same time decrease below $Q_2$ as consumers react to higher petroleum product prices. The gap would therefore shrink with an increase in the price. Such a price increase might be achieved, for example, by a tariff on imported oil if the world price is at $7.88/bbl or below; or by allowing the domestic price to rise toward the world price if it is above $7.88/bbl.

Alternatively, a quota that limits imports to an amount less than the gap between $Q_1$ and $Q_2$ would find consumers unable to obtain the quantity they desire at the prevailing price. Consequently, consumers would be willing to bid up the price to obtain what they want. As the price rises, the quantity demanded falls and the quantity supplied rises until a point is reached where the market is cleared at the allowed level of imports.

The price–quantity relationship of concern here depends on the price elasticities of supply and demand, i.e., the percentage changes in quantity supplied and demanded as a result of a given percentage change in the price. Available estimates of these price elasticities are subject to debate, based as they are on past evidence with no necessary relationship to future supply and demand. Nevertheless, the argument will proceed using the estimates commonly suggested by others: $-0.5$ for demand and 1.0 for supply.[12] That is, a 1.0 percent increase in price will produce a 0.5 percent decrease in quantity demanded and a 1.0 percent increase in quantity supplied, respectively. Readers in disagreement with the assumed values of these parameters are urged to substitute alternative values for comparison with our results. It is suggested, however, that the implications will not be substantially changed.

The estimated relationship between U.S. prices and oil imports is given in table 2-5 for three degrees of restriction imposed on the base case: banning Arab imports, banning non-Western Hemisphere imports, and banning all imports.[13] The two cases of partial import

---

[12]See James C. Burrows and Thomas A. Domencich, *An Analysis of the United States Oil Import Quota* (Lexington, Mass.: Heath Lexington, Inc., 1970) and Stephen L. McDonald, "Incentive Policy and Supplies of Energy Sources," paper presented before the Joint AAEA–AEA Meetings in New York, December 29, 1973. These estimates were also used in the Cabinet Task Force Report on Oil Import Control, *The Oil Import Question* (Washington, U.S. Government Printing Office, 1970).

[13]The estimates were calculated by noting that the change in imports due to change in price ($\Delta M/\Delta P$) is equal to the change in quantity supplied ($Q_s$) and quantity demanded ($Q_d$) due to a change in price, i.e.,

restrictions may be interpreted, perhaps more appropriately, in terms of holding the level of imports to the corresponding proportion of domestic production or consumption. It will be observed that the level of imports falls and the price rises until the limit is reached where domestic production satisfies all domestic consumption. For the midpoint estimate, the self-sufficiency price is $10.24/bbl; for the relatively pessimistic upper estimate, the self-sufficiency price is $12.39/bbl.[14] If only the amount of imports estimated to come from Arab countries is banned, i.e., if imports are restricted by 27 to 30 percent of consumption, the domestic price must rise to $8.43/bbl using the midpoint estimate, or to $10.29/bbl using the upper import estimate.

These estimates imply that domestic self-sufficiency can be attained at a crude oil price not significantly different from recent landed prices for U.S. oil imports. That is, if the U.S. price were allowed to rise to the current price of imports, and expected to remain there, eventually the demand for imports would fall to zero. A number of caveats are in order regarding these estimates. First, the estimates ignore a number of complications that result from the interrelationship between the prices of oil and other energy sources. The elasticity parameters used to obtain the estimates assume that prices of competing goods remain constant. Assuming as we do that natural gas prices are free to vary, an increase in crude oil prices is likely to produce an increase in natural gas prices. Coal prices could be expected to rise as well. This might prevent the demand for crude oil from declining as fast as calculated above. On the other hand, given the wide disparity between price variations experienced in the past and those hypothesized in table 2-5, refinements such as these are of a relatively low order of significance.

Second, the price and quantity estimates derived here are to be regarded as long-run equilibrium values that result after sufficient

---

$$\frac{\Delta M}{\Delta P} = \frac{\Delta Q_s}{\Delta P} + \frac{\Delta Q_d}{\Delta P}$$

Substituting in the definitions for the price elasticities of demand $(E_d)$ and supply $(E_s)$, we can solve for the necessary percentage change in price.

$$\frac{\Delta P}{P} = \frac{\Delta M}{E_s Q_s = E_d Q_d}$$

[14]For comparison with these estimates, if the demand elasticity is cut in half $(-0.25)$, the midpoint equilibrium price is $10.85/bbl and the upper bound equilibrium price is $13.99/bbl. If the elasticity of supply is halved (0.5), the midpoint price is $11.13/bbl and the upper price is $13.80/bbl.

TABLE 2-5.  Estimated U.S. Price–Import Relationship in 1985

| Import reduction | Price[a] ($/bbl) | Domestic production[b] (MMb/day) | Domestic consumption (MMb/day) | Imports (MMb/day) | Imports as a percentage of | |
|---|---|---|---|---|---|---|
| | | | | | Production | Consumption |
| Base case | | | | | | |
| Midpoint estimate | 7.88 | 14.6 | 22.2 | 7.6 | 52.1 | 34.2 |
| Upper estimate | 7.88 | 12.3 | 27.1 | 14.8 | 120.3 | 54.6 |
| Ban Arab imports | | | | | | |
| Midpoint estimate | 8.43 | 15.6 | 21.4 | 5.8 | 37.2 | 27.1 |
| Upper estimate | 10.29 | 16.1 | 23.0 | 6.9 | 42.8 | 30.0 |
| Ban non-W. Hemisphere imports | | | | | | |
| Midpoint estimate | 9.14 | 16.9 | 20.4 | 3.5 | 20.7 | 17.2 |
| Upper estimate | 11.32 | 17.7 | 21.2 | 3.5 | 19.8 | 16.5 |
| Ban all imports | | | | | | |
| Midpoint estimate | 10.24 | 19.0 | 19.0 | 0.0 | 0.0 | 0.0 |
| Upper estimate | 12.39 | 19.3 | 19.3 | 0.0 | 0.0 | 0.0 |

[a]Applying the formula in footnote 13 to a $7.88 base price gives the calculated price at import volumes less than the base case.
[b]The price increases in the first column and the assumed price elasticities of 1.0 for supply and −0.5 for demand.

time has passed to allow for appropriate adjustments in production and consumption. Imposing an immediate ban on all or part of the imports as described earlier would force prices much higher in the short run than those estimated here.

Comparisons between the price estimates in table 2-5 and current world prices require caution as well. Current world prices reflect only temporary market conditions and cannot be expected to form the basis for future production and consumption decisions. Producers and consumers must expect the prices estimated in table 2-5 to prevail indefinitely before they can make the corresponding long-run decisions. World prices, on the other hand, may soon fall to levels below that necessary to support higher cost domestic production and discourage consumption. In this event, the domestic price would have to be guaranteed if the desired level of self-sufficiency is to be maintained. This, of course, is where the policy options of import controls, subsidies, or taxes have their influence.

## IMPORT CONTROLS VERSUS SUBSIDIES AND CONSUMPTION TAXES

Tariffs and quotas have the same effect on domestic production as a subsidy and the same effect on domestic consumption as a consumption tax insofar as the domestic price is increased. The higher price encourages domestic production in the same way as an added subsidy to producers encourages additional output. The higher price discourages consumption in the same way as an added tax decreases purchases. However, a subsidy can be used to increase the revenues to producers without increasing consumer prices (except indirectly through taxation), whereas the consumption tax can be used to increase consumer prices while holding the revenues to producers constant. A subsidy does not distort consumption patterns, and a consumption tax does not distort production incentives. Both have the advantage of showing explicitly the cost of encouraging domestic production or discouraging domestic consumption, in contrast to the hidden costs of quotas and tariffs. Subsidies and taxes (including tariffs) produce their effects by working through prices; with quotas, in contrast, prices become a result rather than a determinant.

The amount of a subsidy or consumption tax required to reduce dependence on imports is estimated using the same 1985 output–consumption estimates and price elasticity assumptions as before. Figure 2-2 compares the production subsidy with import con-

FIGURE 2-2. Using a subsidy to achieve self-sufficiency.

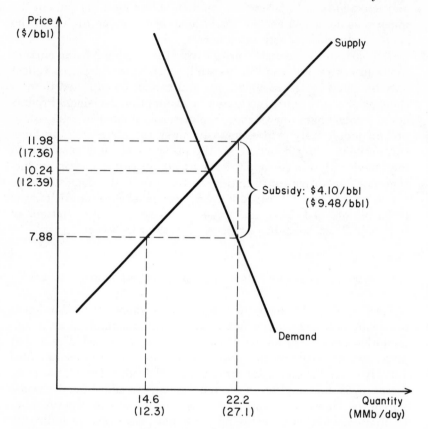

trols in the case of complete self-sufficiency. The values on the quantity axis refer to the midpoint and upper limit (in parentheses) projections at the $7.88/bbl price. As reported above, a price of $10.24 clears the market ($12.39 in the upper limit). If the price to consumers remains fixed at $7.88/bbl, the subsidy must be sufficient to increase domestic production to meet consumption. With the subsidy, the necessary per-unit payment to producers is estimated at $11.98/bbl in the midpoint case and $17.36/bbl in the upper case.

Figure 2-3 compares the tax required to reduce consumption with the amount of output expected at the $7.88/bbl price. Including the tax, it is estimated that the price must rise to $13.28/bbl before

FIGURE 2-3.    Using a consumption tax to achieve self-sufficiency.

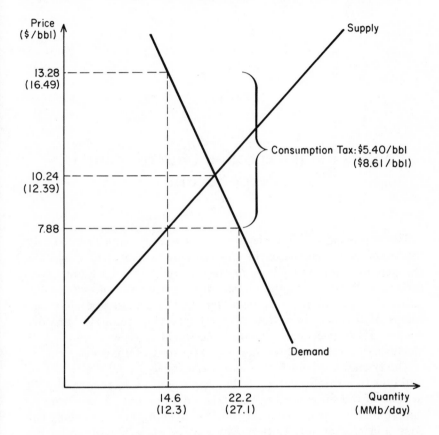

consumption falls to the midpoint supply estimate of 14.6 MMb/day
and to $16.49/bbl to reduce consumption to the 12.3 MMb/day level
of output in the pessimistic case.

These estimates should be regarded as illustrative of the relative
magnitudes involved rather than as precise estimates. They are
predicated on the assumptions stated above and may be altered
easily by substituting other assumptions. Moreover, the effects described refer to subsidies and taxes as general policy tools. Individual subsidy and tax schemes have peculiar characteristics that
produce differential side effects. A discussion of particular policies is
deferred to chapter 5.

# 3

# Policies of Oil-Exporting Countries and U.S. Energy Security

The appropriate U.S. policy to achieve energy security must take into account the policies of the oil-exporting nations. Unfortunately, the policies they will follow in the future cannot be known with certainty. While economic factors are significant, political reality alters revenue-maximizing behavior. In this chapter possible outcomes of the decisions taken by oil-exporting countries are discussed. These decisions structure the world price with which U.S. producers must compete if they are to retain U.S. markets.

The crucial issue for U.S. energy security is the price at which oil will flow in future world trade. That price will be determined by the actions of the oil-exporting countries. A price that arises from an agreement among oil exporters will be much higher in the long run than a price that derives from competitive world trading in energy sources. The oil-exporting nations prefer to control the price of oil and to maintain the consequent higher revenues rather than to accept lower prices and revenues. The necessary conditions for controlling that price are discussed in later sections.

If the oil-exporting countries resolved to administer the price of oil and created a production-limiting mechanism to maintain that price, a cartel would result.[1] Still unanswered would be the question

---

[1]The use of the word "cartel" for the oil-exporting nation organization does not imply complete agreement on the details of price and output regulation, but instead agreement with the principle. The mechanism to achieve this goal need not be explicit. Similarly, the use of the same word with reference to the international oil companies is intended to have no specific legal or theoretical cognate. It simply

of what the cartel price would be. A price must be chosen with the full understanding that its selection is in part arbitrary, and that any price selected will be less satisfactory than some other price for most, if not all, oil exporters. The long-run cartel price in such circumstances cannot be viewed as inviolate over time. Changing conditions, shifts in national power, and other matters associated with the interests of individual member nations will bring changes in the cartel price. Much of the analysis that follows is designed to indicate which nations can be predicted to act in such a way as to bring temporary fluctuations in any cartel price agreed upon.

Fluctuations in the cartel price either can be self-limiting or they can bring about disintegration of the structure that maintains the administered price. The contention developed here is that there are equilibrating forces which will return an errant oil price toward a specified long-run cartel price and restrain the range of oil price fluctuations. Nonetheless, long-term success of the country cartel is not assured. Factors which might threaten that continuity are explored in the latter part of this chapter. The estimated level to which oil prices might fall if the country cartel dissolves is indicated on pages 71–76 because of its importance to U.S. energy producers and energy policy.

## CONTROL OF OIL PRODUCTION AND PRICING

The institutions through which petroleum is supplied to the world underwent great change between 1950 and 1974. In this section we describe the patterns of international industry operations and of industry–government interaction during this period. In summary, the international exporting industry moved from domination by seven major oil companies to a period of bilateral oligopoly (during which the countries controlled production and the companies controlled marketing) to its current state of expanding exporting country dominance. The purposes of this analysis can be achieved without the detail or documentation required in an exhaustive examina-

---

implies that because of actions taken, whether independently or not, the result is a higher price and lower output than would have been achieved if the market had exhibited other organizational patterns. For clarity, the words "country" or "company" will be used with "cartel" when doubt may exist as to the referent.

tion of these developments. A number of sources exist for those who wish to pursue these matters further.[2]

## The Changing Role of the Majors

In 1950, seven oil companies virtually controlled the international oil business. The producing group of which they were members was singularly successful. The saga of the development of their control over production and price is rich in the romance of international intrigue, high finance, great risks, and luck. Unfortunately, the retelling of this story is beyond the scope of this work. The seven major oil companies were British, Dutch, and American by nationality, but their interests and loyalties were international. Each has gone through name changes, but by their present designation the companies were: British Petroleum, a British company with substantial government ownership; Royal Dutch/Shell, an Anglo-Dutch firm; and Mobil, Exxon, Socal, Texaco, and Gulf of the United States. A small French company with less influence at least deserves mention—Compagnie Française des Petroles (CFP).

These companies controlled virtually all of the oil reserves in the underdeveloped oil-exporting regions of the world. They exercised joint maximizing behavior by programming production to reduce excess supplies and by honoring exclusive marketing areas. Production programs were implicitly suggested and explicitly agreed upon; information on company plans and policies was transmitted through the inevitable interchanges within the maze of interlocking joint

---

[2]There are a number of excellent studies of different aspects of the international petroleum market which cover the period in which we are interested. Among the generally available sources, the following can be especially recommended: M. A. Adelman, *The World Petroleum Market* (Baltimore: Johns Hopkins University Press for Resources for the Future, 1972); Joel Darmstadter and others, *Energy in the World Economy* (Baltimore: Johns Hopkins University Press for Resources for the Future, 1971); J. E. Hartshorn, *Oil Companies and Governments* (London: Faber and Faber, 1962); Stephen Hemsley Longrigg, *Oil in the Middle East* 3rd ed. (New York: Oxford University Press, 1968); Zuhayr Mikdashi, *A Financial Analysis of Middle Eastern Oil Concessions: 1901–1965* (New York: Praeger, 1966); Zuhayr Mikdashi, *The Community of Oil Exporting Countries* (Ithaca: Cornell University Press, 1972); Sam H. Schurr and Paul T. Homan, *Middle Eastern Oil and the Western World* (New York: American Elsevier, 1971); Federal Trade Commission, *The International Petroleum Cartel* (Washington: U.S. Government Printing Office, 1952).

Periodical sources are invaluable aids in understanding this material. Among the most valuable periodicals consulted were: *The Middle East Economic Survey, The Oil and Gas Journal, Petroleum Economist, Petroleum Intelligence Weekly,* and *Petroleum Times.*

ventures in concession areas. This control over the resource base made it possible for the companies to avoid the overproduction that would have tempted price competition in world markets. It also restricted the bargaining power of the host governments who faced unified or single bidders for concessions. Further, as Iran found out later, even nationalized petroleum could not be produced without the aid of the majors because it could not be sold without their cooperation. The majors controlled the demand for crude oil because they controlled the transportation, processing, and marketing of its products.

A number of factors helped to undermine the role of the major oil companies. The ultimate failure of company control, however, lay in its being unable to foreclose entry into the prolific Middle Eastern and North African oil-producing regions. Two conditions were necessary to effect entry into the foreign oil provinces: markets for products and concessions. In the postwar period a number of sizable U.S. domestic firms, and a number of national companies in other countries as well, were unable to produce as much crude as they could process and market. They had the downstream outlets for foreign crude. The great Greek tanker magnates supplied the vital transport links between crude sources and refining and consuming regions. A change in the times provided the concessions. The decline of colonialism (and U.S. policy) left the Middle East and Africa less controlled by the home governments of the oil companies, which had previously supported the cartelization of the oil business. Change came even in the potential oil countries that retained a traditional monarchy. New sources of revenue were required, and new companies were given access to prospective producing regions.

The enlarged output from new concessions began reaching the world market in the early 1950s. The effect it had in the United States was significant for the future role of the international majors. So long as imports into the United States were controlled by the international majors, these exports were restricted to a level which, while replacing some U.S. output growth, did not bring price weakness. On the other hand, when smaller U.S. firms obtained low-cost foreign oil, they had no individual long-run interest in maintaining market stability. They marketed as much oil as they could. Expanded oil imports struck the U.S. market in late 1955. Imports rose from 239 million barrels in 1954 to 285 million in 1955, 342 million in 1956, and 373 million in 1957. Imports would have been higher still in 1957 except for the first Suez Crisis. The 7 "traditional" importers became a group of 36 in 1957 and by February of 1958 numbered 57. The result was oil import controls to protect domestic producers in 1957. At first the controls were voluntary, apparently on the faulty assumption that the oil importers themselves would discipline the market if rules were established. The mandatory controls which followed in 1959 favored existing importers by granting "grandfather" rights to special allocations based on import history. These controls were a blow to the U.S. independents who had recently moved abroad, but these firms were sufficiently well established that most were able to survive.

The imposition of U.S. import controls had two effects on the international oil market: it somewhat restricted the market for exported oil, and it handicapped new entrants in their efforts to initiate or expand output. Oil-exporting countries were harmed by both effects. Their output ceased to grow at the expected rate, and they had less bargaining power with which to improve concession terms. It is not surprising that the major oil-exporting countries chose this time to create an organization to press their claims. Cuts in posted price were the proximate cause. The Organization of Petroleum Exporting Countries (OPEC) was founded in September 1960, following the second unilateral revision of posted prices by the oil companies in August 1960.

## Ascending Power of the Oil-Exporting Countries

The later shift in bargaining power between OPEC and the oil companies came partly because of the skill and cohesiveness of OPEC but primarily because of erosion of the power of the international oil firms. It is difficult to identify independent companies because of corporate interlocks and changes of identity, but the following figures are at least suggestive of the increase in the number of firms in the Middle East alone. In 1940 the seven majors and CFP were the only significant producers in the Middle East. By 1950 ten U.S. independents were present, though their production was miniscule. By 1955 seven more U.S. firms were active, the majority of them brought in through the Iranian Consortium. Two more U.S. firms and two Japanese companies had entered by 1960, to be joined in the next 5 years by eight more U.S. firms and seven other foreign companies. Between 1965 and 1970, 31 companies entered, leaving (after mergers) the 7 majors, 24 U.S. independents, 31 foreign companies, and 13 government entities, for a total of 75 firms. The majors continued, of course, to produce and market the bulk of the oil.

Further deterioration of posted prices did not occur in the 1960s, though rising import prices brought declining terms of trade for oil-exporting nations. A number of improvements in contract terms were negotiated by the exporters. Starting in 1962, royalty was no longer treated as a part of the income tax owed under the 50-50 profit split; it was a deduction from income, and hence the companies were required to pay a royalty in addition to taxes. Marketing allowances were phased out. By the end of the decade, country participation in the operation and management of the concessions had become a declared goal. Employment of natives in positions of responsibility was obtained. These changes came because the new entrants brought competition for concessions. They needed more oil to meet their expected commitments. More importantly, the new firms were vulnerable. Country recalcitrance, a mere annoyance to a major with alternative supply sources, was a desperate threat to a newcomer. One by one, more onerous conditions were forced on the independents, and then used as a basis for negotiation with the majors.

These changes meant that the atmosphere of country–company relations was reversed in the 10 years following the 1959 unilateral reduction in posted price. By 1969–70, the market structure could best be described as one of bilateral oligopoly. There was a degree of cooperation between both sides to retain a world price above that which would hold if competitive conditions existed in all markets. The relative disposition of the total monopoly profit between the countries and companies, however, was uneasy and unstable. There was no clear focal point for division of the monopoly returns. The balance began swinging toward the OPEC countries in 1970. Then market conditions, good fortune, and a risky but successful venture by OPEC started it toward the dominant status it enjoyed by 1974.

A number of contributory conditions left the world with a tighter market for oil in 1970 than had previously been the case (see chapter 1). Taking advantage of these conditions, Libya unilaterally curtailed crude production in an effort to force higher taxes and a higher posted price on the oil companies. Higher posted prices and tax rates for African and Mediterranian crudes followed on September 1, 1970, breaking a long period of price stability. The per barrel revenues of the Persian Gulf and Venezuelan governments were increased before the end of that same year. It soon appeared that any settlement with one country served as the basis for reopened negotiations elsewhere. In response, the international firms initiated new talks in an effort to reassert their former bargaining position and to stabilize world prices.

To redress the balance of bargaining power, the companies revised their strategy to one of industry-wide bargaining. They thought that in this way the weakest country's need for a settlement would discipline the OPEC group as a whole. At the very least, contract sanctity, which had virtually disappeared in the frantic price leapfrogging in the latter part of 1970, would be reasserted. While the companies wanted joint negotiations over all OPEC oil, the exporting countries wanted the agreements broken into separate phases, the first covering the Persian Gulf. A stalemate resulted in the early 1971 talks in Teheran.

With the talks recessed, the crucial gamble was taken by OPEC. In a special meeting of OPEC representatives on February 3, 1971, a resolution was adopted that pledged every country to enact (on February 15) the necessary legal or legislative measures to embargo (on February 21) shipments of crude and products to *any company* from *any source,* if that company did not meet the OPEC minimum demands for the Persian Gulf producing area. The oil companies, perhaps at the urging of their own governments, elected to meet the OPEC terms. In April an agreement was reached in Tripoli that was the basis for the new prices for Mediterranean and African crudes. Hence, in response to a threat from the oil companies, the OPEC countries were forced to take a public position. From that time onward, the oil exporters could only stand firm and win, or else experience a dramatic reversal in their hitherto slow progress toward equality in bargaining strength. Perhaps the size of the stake OPEC laid on the table was a crucial element in convincing the oil companies that only through a long and bruising battle, if then, could the OPEC position be reversed. The immediate effect of Teheran and Tripoli was to raise the revenues of the OPEC countries; the long-run effect was to validate the OPEC cartel.

Rising world energy demand, when accompanied by the stronger bargaining position of a newly vitalized OPEC, brought increasing efforts for higher prices and better terms for host countries almost as soon as the Teheran and Tripoli agreements were signed. The participants in the Teheran and Tripoli agreements met in Geneva and agreed to an increase in posted prices of 8.49 percent effective January 20, 1972, to compensate for the decline in the value of the dollar. The U.S. devaluation of 1973 resulted in an additional 11.9 percent increase in the posted prices; the parties further agreed to an automatic adjustment of dollar postings referenced to an index based on major currencies. In late 1972 the major producing countries of the Middle East negotiated participation pacts calling for the transfer of interest in the operating companies to the host countries upon payment of compensation. Nigeria accepted a participation agreement with an immediate 35 percent share in April of 1973. Other OPEC nations set up national oil companies to explore for and develop reserves *de novo,* while some nations chose partial or total nationalization as a means of achieving control over their oil resource. The apparent willingness of OPEC countries to stand behind each other in enforcing these demands made success possible. Clearly, the relative position of the host countries improved steadily after 1971.

### OPEC After the October War

The final stage of the development of company–OPEC relations followed the October 1973 war. Two changes mark this episode. First, the OPEC countries abandoned all pretense at negotiation with the oil companies. They instead unilaterally altered prices and production rates. Second, the countries themselves started marketing significant quantities of oil. By the end of 1973 there was no question that the oil-exporting countries were making virtually all of the important decisions affecting the crude oil production phase of world oil trade.

True sovereignty now rests with the exporting countries. Increasingly they own the crude oil being produced. The concessionaire companies may influence, but they do not control, host country decisions. In the important matter of production levels, the countries can mandate increases or decreases if they are willing to accept the economic and political consequences of the changes. Their ability to restrict production is direct; incentive and sanction mechanisms to increase production also exist.

The fact that the countries control their output and marketing decisions does not imply, of course, that an effective producing country cartel exists. The fact that oil prices are well above the long-run competitive price does not prove the existence of a cartel either. In periods of rapid shifts in demand and supply, where long lead times are required for response and where short-run elasticities are low, prices well above the long-run competitive equilibrium are quite consistent with either a competitive or a cartelized market. The establishment of a cartel requires the ability to control output to maintain a price above the competitive equilibrium over time, and the ability to adjust to shifts both in external factors and in relative positions of members of the group. In the next section these requirements will be examined as they apply to OPEC and the special problem of a cartel whose members are also sovereign nations.

## OPEC as a Producer Cartel

This section attempts to describe the policies that the oil-exporting countries must follow if they are to act as a cartel and to indicate the constraints within which they must work. The problem is complex. Our approach is to formulate a model based on a reading of the goals of each of the member states of OPEC.[3] While political and historical factors play some part in the positions taken by each country, much can be explained by reference to country self-interest expressed in economic terms.

The policies which OPEC must pursue if it is to succeed follow from the usual analysis of cartel behavior. It must restrict production to raise prices. OPEC's ability to follow usual cartel policies, however, is constrained by the special circumstance that it is made up of members with very diverse interests over which no supreme power either exists or is likely to be created. Hence, success for OPEC implies that a price–output policy can be formulated to which countries will voluntarily adhere. This means that a formula must be found which meets at least the minimum goals of each country, and makes each better off than it would be if it acted competitively. In essence, a reference price at which petroleum will be sold must be chosen, along with the quantity that each country will be allowed to sell.

The first task facing OPEC is the selection of an overall price–output policy for oil-exporting countries. In order to maximize the current value of its holdings, the cartel must choose a policy that

---

[3]The OPEC countries we deal with here are eleven in number: Algeria, Indonesia, Iran, Iraq, Kuwait, Libya, Nigeria, Qatar, Saudi Arabia, United Arab Emirates (formerly Abu Dhabi, including the former Trucial States, of which only Duabi is an important producer), and Venezuela. Ecuador has been granted membership (1974) but is not included in this discussion.

will equate the marginal cost of producing a barrel of oil now with the extra revenue it would receive today from producing that oil. In simple terms, the decision problem for OPEC taken as a whole can be indicated by the familiar monopoly model, with marginal cost interpreted as including direct costs of production *and* the current value of the income forgone in the future because a barrel of petroleum is produced now. Figure 3-1 depicts this situation on the assumption that the cartel maximizes joint profits. A constant marginal cost function is used because the most significant element in it, the present value of the revenue from a barrel of petroleum produced in the future, is invariant with output.[4] Two marginal cost curves are drawn, "MC-Cartel," reflecting an expectation that the cartel price will continue to exist in the long run; and "MC-Competitive," reflecting an expectation that a competitive market will prevail because the cartel falls.[5] The relationship shown in figure 3-1 is important for what it tells us about the factors that influence the price–output preference of OPEC. A high expected future price from maintenance of the cartel would raise marginal cost and result in lower current output rates and higher current prices. The current return from cartel action is reflected in the gap between the cartel marginal cost and price. The current return from expected future cartel behavior is reflected in the difference between the cartel marginal cost and the competitive marginal cost. If price expectations fell, in order to maximize revenue the cartel would lower current prices and increase output.

Joint action to restrict production enables the OPEC countries to receive monopoly profits from their reserves. Maintenance of the cartel is, then, in the best interest of the participants taken as a whole.[6] The primary cost of these benefits to the individual cartel

---

[4] Moreover, not only are production costs small compared with the asset value of the oil produced, but there is no reason to believe that production costs would rise significantly with output levels within a considerable range. A more complex assumption as to the shape of the marginal cost curve would complicate the analysis but not alter our conclusions.

[5] An approach to the selection of the future cartel price from which this marginal cost is derived is presented in the next section.

[6] Maintenance of the cartel is also in the interest of all other producers of energy competitive with OPEC oil—and going further, of benefit to producers of substitute goods. To the extent that these competitors to OPEC produce energy in response to the higher cartel price, they reduce the demand for OPEC oil, and make even more restriction of OPEC production necessary. It is thus to OPEC's advantage to bring competitive producers within its ambit so that they bear part of the burden of restrict-

FIGURE 3-1. Cartel and competitive pricing strategy.

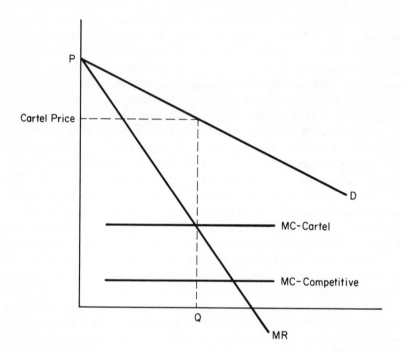

member is the return from sales forgone at prices exceeding marginal costs. The question the cartel must resolve is the allocation of
these "costs" among its members in the form of production restrictions. None of the truly efficient means of organizing a cartel are
open to OPEC. It cannot rationalize production to maximize the
summed net revenues of the group through increased efficiency. It
can only distribute production quotas among member countries,
hoping that members will not sabotage joint goals by expanding
production beyond those limits.

Enforcement of cartel policy is made easier by several factors
which may be suggested here.[7] The trail of broken contracts left by
OPEC members has destroyed the credibility of long-term commit-

---

ing production. It is similarly desirable for those outside the cartel to stay outside,
where they may reap the benefits of OPEC production control without sharing any of
its costs.

[7] A final evaluation of the likelihood that an OPEC cartel will be sustained is
delayed to a later section.

ments by purchasers as well. Consequently, the ability of one member to gain a long-run advantage by undercutting the cartel and gaining a *permanent* larger market share is reduced. Retaliatory price cutting from other members could be expected, unless terms remain secret. Purchasers could be predicted then to abrogate the contract with the initial price cutter to obtain even cheaper supplies. Consequently, the initial price cutting is less likely to take place. Second, the cartel *qua* cartel, lacking sovereignity, must adopt a policy that would make each member country better off than it would be in the absence of the cartel. While the actual price–output policy chosen is unlikely to maximize the summed net revenues of the members, it will necessarily be better for each country than the competitive alternative. All will suffer if the cartel collapses. Finally, negotiating room exists within the cartel. While intracartel tensions may arise, they can lead to adjustments to cartel price–output policy and to changed market shares which again make each country better off than it would be if it initiated price cutting. Consequently, it is quite possible that, even without an explicit enforcement mechanism, member countries of OPEC will restrain output in such a fashion as to maintain a price well above the competitive level. Fluctuations in price are inherent in the process by which competition is restrained, but those fluctuations need not lead to the unrestrained price cutting that would dissolve the cartel. Before the limits to those fluctuations can be described, it is necessary to analyze the process by which the long-run cartel price itself is chosen.

## The Long-Run Cartel Price

There is no way for the oil-exporting countries to determine what the optimal price–output policy for OPEC would be if maximum joint revenue were the OPEC goal. Uncertainty about the future, when added to the ignorance of the present, makes it impossible to select any policy as unambiguously optimal in these terms. Moreover, even if such a policy could be identified, it would leave some countries worse off than would alternative policies. Hence, the members of OPEC cannot identify an "optimal" policy, and even if they could, some would not choose it. This is the nature of the problem facing sovereign entities.

The problem, then, is to predict which OPEC price–output policy, among the many possible, will evolve, if indeed any agreement is reached at all. Given the uncertainty surrounding identification of an optimal policy, it is the thesis of this section that OPEC will be forced to choose an "obvious" policy. The analytical construct of a focal point in conditions of decision making that are disorganized or disoriented is relied upon here. This concept, as applied to quasi-economic decision making, was developed by Thomas C. Schelling and others.[8] It follows and is underpinned by the literature in social psychology that derives from the analysis of group behavior pioneered by Musafer Sherif.

In selecting among possible "focal points," a number of reasons lead to the conclusion that the U.S. self-sufficiency price will be "obvious," and hence the target price for the OPEC cartel. The U.S. energy market is the largest in the world, consuming one-third of the world's energy. The major international oil companies are head-quartered in the United States. The United States is among the world's largest oil producers and its energy-intensive exports compete throughout the world. It is the center of energy research. Its economy is the largest in the world, and it has the world's most energy-intensive production and consumption patterns. The price of U.S. energy must be a major factor in the thought processes of OPEC decision-makers.

The declaration of Project Independence is a further reason to believe that the U.S. self-sufficiency price will become the OPEC focal price. Project Independence signals the OPEC countries that great expansion in the U.S. market is not to be expected, whatever prices are selected. They cannot, as a group, reduce prices below the U.S. self-sufficiency price and thereby markedly expand their sales

---

[8]Thomas C. Schelling, *The Strategy of Conflict* (Cambridge: Harvard University Press, 1963), pp. 54–80. The body of work derived from Schelling's original formulation demonstrates that under conditions of uncertainty, where collective behavior is required, most often unilateral decisions will be made to adopt the most *obvious* choice. Uncertainty can arise from lack of communication among members of the collective. In this case, the obvious choice will be a point distinct from other possible choices; a unique point. When communication exists among cooperating parties, uncertainty can still exist if there is no clearly correct choice. Again, any choice which is unique will likely be chosen. Finally, even among hostile parties caught up in a zero sum game which demands an outcome, a resolution which each can identify as having some exogenously derived uniqueness or "rightness" will most likely be the choice on which all coalesce. In each situation where there is only one alternative visible to all parties, Schelling suggests that the outcome is determinable. He terms the outcome on which all parties converge the "focal point." In our situation, OPEC members seek a "focal price" on which cartel production policy can be based.

in the United States. In the short run those quantities will be sold in any event, and hence there is no reason to go below the self-sufficiency price. To protect long-term markets for the future, a higher price, which would induce excess capacity in the U.S., would be unwise. Project Independence tells OPEC something else as well. Energy costs in the United States are going to be quite high in the future. Insofar, then, as other countries' energy-intensive goods are competitive with those of the United States, those countries will not be placed at a disadvantage if energy prices to them are equal to those in the U.S. market. Moreover, autarchic U.S. energy prices are likely to be at the lower limit of the entry prices required by other countries to achieve significant increases in their own energy self-sufficiency. Hence, OPEC need not reduce prices below the U.S. self-sufficiency price to maintain its non-U.S. markets. On the other hand, an OPEC price above the U.S. self-sufficiency price in the long run would bring U.S. energy exports into the world market. More importantly, it could make some U.S. energy technology attractive to other consumers and threaten long-term OPEC markets.

A final point may be made in support of the U.S. self-sufficiency price as the focus for cartel behavior. The explicit public philosophy of leaders of oil-exporting nations has been that such countries should receive the full value of the oil they export. That value, they argue, is measured by the cost of substitutes. In making this concept operative, OPEC and some of its member governments have commissioned private studies to determine what the cost of substitutes for crude oil might be. It is clearly in their interest to price oil so that such substitutes, essential for U.S. self-sufficiency, remain only marginally economic until the bulk of the natural crude is produced.

In chapter 2 a range of prices was posited that would yield long-run self-sufficiency in the United States under different assumptions. The thesis here is that OPEC will select its target cartel price from within that range. Whether that price proves to be the actual self-sufficiency price is of little importance. Developments would not be greatly affected if OPEC selected a price somewhat above or somewhat below that level. What is crucial is that some focal price based on estimates of the U.S. self-sufficiency price be chosen by OPEC as its long-run target cartel price. For the purposes of analysis and presentation here, it is assumed that OPEC decision-makers will coalesce around a price of $10 per barrel (1973 prices).

The upper bound of the potential OPEC cartel price, freight on board (f.o.b.) the member country, would be the netback from the

TABLE 3-1.   Projected Long-Run Cartel Price f.o.b.
Each OPEC Country

| OPEC Country | Transport cost[a] (1) | Representatives crude[b] | | | Netback factor (Col. 1 + Col. 4) (5) | f.o.b. Price ($10 + Col. 5) (6) |
|---|---|---|---|---|---|---|
| | | Gravity (degrees) (2) | Sulfur (%) (3) | Quality differential (4) | | |
| Algeria | $0.60 | 44 | 0.1 | $ 0.88 | $+0.28 | $10.28 |
| Indonesia | 0.91 | 35 | 0.1 | 0.34 | −0.57 | 9.43 |
| Iran | 1.50 | 33 | 1.5 | −0.73 | −2.23 | 7.77 |
| Iraq | 1.50 | 36 | 2.0 | −0.93 | −2.43 | 7.57 |
| Kuwait | 1.50 | 31 | 2.5 | −1.49 | −2.99 | 7.01 |
| Libya | 0.60 | 39 | 0.2 | 0.51 | −0.09 | 9.91 |
| Nigeria | 0.60 | 33 | 0.2 | 0.18 | −0.42 | 9.58 |
| Qatar | 1.50 | 39 | 1.3 | −0.26 | −1.76 | 8.24 |
| Saudi Arabia | 1.50 | 34 | 2.0 | −1.05 | −2.55 | 7.45 |
| United Arab Emirates | 1.50 | 39 | 0.7 | 0.16 | −1.34 | 8.66 |
| Venezuela | 0.25 | 29 | 1.7 | −0.99 | −1.24 | 8.76 |

[a]Rounded, at world scale 100, as a reasonable long-run approximation.
[b]The marker crude is taken to be 0.5 percent sulfur and 34°. The sulfur differential is calculated as $0.07/bbl for each 0.1 percent above or below 0.5 percent sulfur. The gravity differential is set at $0.06/bbl added for all degrees above 34°; $0.03 subtracted for all below. ("Representative Crude Oil Prices," The Petroleum Economist, April 1974, p. 58.) The adjustments noted are perhaps overstated because they are based on the posted price. Given the confusion in actual prices, and the fact that some oil moved at the posted price (or 93 percent of it) during the time to which these differentials applied, the use of the full differential appeared appropriate. Representative crudes were selected for each country with no formal attempt at weighting the different crudes by prospective production. Data on crude types were taken from: "Analysis of the World's Crudes," and "Major Oil Fields Around the World," The International Petroleum Encyclopedia 1973, Tulsa, Petroleum Publishing Co., 1973, pp. 426–433; 198–204.

U.S. price. The per barrel revenue to each country would be $10 *minus* transportation costs and quality differentials as per the U.S. market.[9] Table 3-1 summarizes the factors leading to the expected per barrel revenue (column 6) of the OPEC countries at the posited

[9]Note that the f.o.b. price to each country is not a measure of net revenue but of gross revenue per barrel. Production costs (for the most part insignificant as a proportion of the total) and a margin to cover transaction costs between the exporting country and the United States must be subtracted to reach a net revenue figure. The f.o.b. price *can* be taken to represent revenues actually flowing to the OPEC country, however, because in this model the country is assumed to control and pay for production and other services required.

cost, insurance, and freight (c.i.f.) price of $10. This price represents the revenues they can each expect to receive per barrel if the cartel restrains output so that the $10 U.S. price can be maintained.

## Variations from the Long-Run Cartel Price

The OPEC output program depends on the cartel price it chooses. Whether the chosen price can be maintained, however, depends upon whether or not the sovereign members of the cartel follow that program. Their actions depend in part on whether the cartel price, and their level of sales, are perceived by them to be appropriate. If all members are satisfied, the price will remain constant. When there is disparity between actual and hoped-for results, variations in the world price may occur. Because the focal price in the long run is the *upper bound* price, and because that price in the long run cannot be increased by any cartel member (except possibly Saudi Arabia) without intolerable cost to the country restricting output, we can ignore upward revisions.[10] Are there countries which would be willing to increase output even if a lower price resulted? If they do exist, which countries might undercut the cartel price? How far might the price cutting go? Note that a natural limit exists for price variations within the cartel. Even if it wanted to sell more oil, no OPEC member would be willing to sell oil at a price below what it conceives to be the present value of its expected future revenue from selling that same oil. Given the f.o.b. price calculated in table 3-1, that present value can be determined by applying the appropriate discount rate and time period.

Variations in the cartel price will occur if one of the member nations of OPEC prefers higher present revenues from oil, and believes it can attain them by increasing output beyond its share of the OPEC market. In essence, such an OPEC member would prefer a redistribution of its oil revenues from the future toward the present, even at the cost of lower overall revenues from its oil deposits over their life. Such an OPEC member would have a higher discount rate than the OPEC average. Determination of relative discount rates for member countries thus identifies those countries which can be

---

[10]The recent period in the petroleum market in which there are calls for higher prices even though excess supply is appearing is an aberration. It bears the seeds of its own correction.

expected to press for higher revenues through higher output and, if their physical production capability permits, put downward pressure on the OPEC cartel price. The process by which Indonesia, Iran, and Nigeria were found to be in this category, and Kuwait, Qatar, Saudi Arabia, and the United Arab Emirates in the other, is discussed below. The reasons that the other countries cannot be placed at either extreme are also expressed. Clearly, the preference of a country for additional current revenue is affected by its situation. Hence country positions relative to the OPEC average will depend upon circumstances *and* revenues. Those positions consequently may change over time.

### Member Country Discount Rates

Determination of the true appropriate internal discount rates for OPEC countries is, of course, very difficult. Happily, precision is not required in an analysis that has as its goal illuminating the forces that will cause OPEC member countries to tend to behave in particular ways. A number of factors working through both political and economic mechanisms influence the discount rate used by a nation's decision-makers. Many of these factors are difficult to measure, and even if measured, more difficult still to weight in estimating the discount rates. Explicit identification of such elements, however, yields useful information. The purpose of this section is to replace the usual vague generalizations about potential member country behavior with a series of variables that form the building blocks from which more carefully reasoned conclusions can be drawn. After considering a number of potential variables, six were included here. These factors are, in no particular order of importance, per capita income, government stability, import demand coverage by foreign exchange holdings, capital absorption capacity, consumption capacity, and petroleum reserve position.

The factors were scaled among the eleven OPEC nations to summarize their effect on the discount rate of each OPEC country. Information was not sufficient for some placements, and in such cases no placement was made. In some cases at the margin, however, there were sufficiently large breaks in the basic data array to make placement decisions with some confidence. This entire procedure admittedly is judgmental and subjective. It is defensible because the explicit identification of the choices made allows those observers with contrary evaluations to identify the source of their differences.

The issues may thus be brought more rapidly to the surface for consideration. This result is not achieved when less explicit (and for that reason perhaps less controversial) judgments are made. The evaluation of each of the six factors influencing member country discount rates is presented in table 3-2.

*Per Capita Income.* Hypothesis: the higher the per capita income, the lower the discount rate.[11] Higher income reduces the compelling necessity for additional current income and, at the same time, increases the relative importance of income continuity for the future. This thesis suggests an absolute income standard for consumption. This result follows because governments have a desire to stay in power. Absolute levels of income are positively correlated with such continuity, and thus those countries with high levels of income will have less political or economic need for more revenue in the present—and conversely more need for it in the future when oil revenues otherwise would begin to decline. The same result follows if the decision-makers are selflessly interested in the well-being of their nation's population. Data problems make a precise determination of per capita income very difficult. Column 1 of table 3-2 was derived from available measures. The United Arab Emirates, for example, with one of the highest per capita income levels in the world, is given an "L" rating.

*Government Stability.* Hypothesis: the more stable the government, the lower the discount rate. Unstable governments need additional revenue in the present to remedy the disaffection of their populations and to purchase arms and the allegiance of armies and police to protect them from internal dissension. Stable governments, on the other hand, can afford to look to the longer run. There is no unambiguous method of scaling governments on the basis of their stability. Nevertheless, the internal condition of the country, the apparent stability of government institutions, and the marked presence or absence of localized disruptions all lend some credence to the evaluations presented in column 2 of the table.

*Foreign Exchange Holdings.* Hypothesis: the lower the ratio of annual imports to foreign exchange holdings, the lower the discount rate. High levels of foreign exchange holdings protect the country against short-run trade difficulties and against temporary inimical

---

[11]The distribution of income may be even more important than its absolute level in affecting this variable. Reliable income distribution figures for the OPEC countries, unfortunately, are unobtainable.

TABLE 3-2.   Ordering of Factors Affecting the Internal Discount Rates for OPEC Countries and Discount Rates Selected[a]

| OPEC Country | Per capita income (1) | Government stability (2) | Foreign exchange holding (3) | Capital absorption capacity (4) | Consumption capacity (5) | Reserve position (1974) (6) | Overall evaluation[c] (7) | Discount Rate selected (percent) (8) |
|---|---|---|---|---|---|---|---|---|
| Algeria | H | — | L | H | H | L | — | 7 |
| Indonesia | H | H | H | H | H | L | H | 12 |
| Iran | H | L | H | H | H | L | H | 10 |
| Iraq | H | H | L | — | L | H | — | 7 |
| Kuwait | L | L | L | L | L | H | L | 3 |
| Libya | L | H | H | — | — | H | — | 7 |
| Nigeria | H | H | L | H | H | L | H | 10 |
| Qatar | L | L | L | L | L | — | L | 2 |
| Saudi Arabia | L | L | L | L | H | H | L | 3 |
| United Arab Emirates | L | L | L | H | L | H | L | 2 |
| Venezuela | L | L | H | H | L | L | — | 7 |

Columns (3)–(6) are grouped under the heading "Factors affecting discount rate[b]".

[a]The bases for the conclusions reported in this table are found in the text.

[b]"L" denotes that the proxy for the country would fall in the lower portion of the array of preferences for discount rates ranked from the lowest to the highest discount rate; "H" means it would fall in the upper portion. Where the cell is empty, no judgment was made.

[c]"L" means that the country on the whole is judged to prefer a higher price–lower output policy than the median OPEC preference; an "H" means the opposite, and an empty cell implies an intermediate choice, not indeterminacy.

economic movements. The country can avoid expansion of petroleum exports if that expansion is judged to have longer-run negative economic consequences. High levels of foreign exchange holdings allow a country to use a lower discount rate in exploiting its reserves. The lack of recent data on OPEC imports complicates determination of the relative position of the countries. The latest data on foreign reserve holdings (the end of 1973) were compared with imports for 1972. The use of 1973 foreign exchange holdings with 1972 import data better represents the prospective status of OPEC countries than would consistent 1972 data. With the sharply higher petroleum prices of 1974, import coverage of foreign exchange holdings will become all but irrelevant for most of the countries of the Middle East and North Africa, with the possible exception of Iran. For Venezuela, Nigeria, and Indonesia, foreign exchange will remain a constraint on policy. While short-term factors caused Kuwait to fall toward the "high" side of the array in the underlying data, it has been given an "L" in column 3.

*Capital Absorption Capacity.* Hypothesis: the greater the capital absorptive capacity, the higher the discount rate. Capital absorption capacity refers to the ability of a country to convert financial assets into domestic productive assets. In effect, this factor measures the ability to achieve economic growth from oil revenues through internal investment. Growth can be achieved directly through the development of natural resources and through industrial and commercial activities or indirectly through acquiring the social infrastructure needed to support economic activity. Oil revenues not absorbed in investment must be used for domestic consumption or for acquisition of foreign assets. The latter alternative is a poor substitute for domestic investment because, among other reasons, foreign investments are hostage to oil export policy.[12] Determining relative absorption capacity among the oil-exporting countries is difficult. It depends on potential as well as actual performance criteria, and as much on social characteristics as on economic ones. An amalgam of the following criteria was used to make the judgments found in

---

[12]Note, for example, the argument proferred by Abd al-Rahman al-'Atiqi in his defense of a 60–40 participation agreement as against nationalization of the Kuwait oil industry. He said, Kuwait should "not take any measures against foreign interests in Kuwait which might be 'imitated to its detriment' in the future by the host countries of Kuwait's own foreign investments." *Middle East Economic Survey,* April 26, 1974, p. 1.

column 4: diversification of the present economy as reflected in the distribution of gross national product by producing sectors, recent trends in the rate of capital formation, relative depth of the local financial market, variety of exports, and literacy rate.

*Consumption Capacity.* Hypothesis: the greater the consumption capacity, the higher the discount rate. Consumption capacity refers to the government's willingness or need to use oil revenues for domestic consumption purposes. Consumption expenditures include use of oil revenues for military forces to increase national security or to preserve the government in power. In that same vein, an increase in personal consumption may be required to reduce the threat of local unrest caused by frustrated aspirations for higher income. Development literature, as well as sociological analysis of revolution and social movements, attests to the importance of unmet aspirations as a source of community disruption. These aspirations may be engendered by comparison with observed conditions over time or by a specific population comparing its status with that of others in its reference group. Increasingly, the reference group of the OPEC countries will be other oil-exporting nations. Even if as a whole their incomes rise, those countries with the lower incomes are likely to remain frustrated. In scaling consumption capacity, both cross-section and time series measures were used as proxies for unmet aspirations. For the time series element, the growth rate in per capita income from 1965 to 1972 was considered. For the cross-section measure, the ratios were formed using as a numerator subject country per capita income less the weighted average of per capita incomes for selected reference countries; the denominator was subject country per capita income. The results from this exercise were modified by expected military expenditures and consideration of absolute standards of living to obtain the results reported in column 5 of table 3-2. An attempt was made to adjust for the timing of the original inflow of oil revenues; hence, the conclusions reported seek to reflect long-run factors.

*Petroleum Reserve Position.* Hypothesis: the petroleum reserve position of a country (relative to its level of production and total OPEC reserves) affects its discount rate, but not unidimensionally. The reserve to production (R/P) ratios of the OPEC members can be divided roughly into two groups, with Kuwait, Saudi Arabia, the United Arab Emirates, Iraq, and Qatar (in descending order) having very high ratios and Venezuela, Algeria, Indonesia, Nigeria, Iran, and Libya (in ascending order) having relatively low ratios.[13] A higher ratio implies a time horizon longer for production from the country's reserves than from OPEC reserves in general. If the production period is held constant, this relationship automatically implies a higher discount rate for a high R/P country because it will wish to increase its production ratio to meet that of other OPEC countries. Additionally, countries with a high R/P ratio may wish to adopt a price policy to protect future markets; a high current price may have a significant effect in reducing future sales. Again, discount rates are *positively* related to R/P ratios. On similar grounds, they need not be. If a country is in a position to convert its petroleum asset quickly, it may choose to produce rapidly before long-term supply and demand shifts can take place. Hence, a country with a low R/P ratio may still have a high discount rate. Finally, whatever the direction of the effect of the R/P ratio, it must be much larger in countries with larger reserves than in those with smaller reserves. Countries with very large absolute reserves must act somewhat "responsibly" because of the effect their actions have on the total market. They must reveal their true preferences more clearly and act upon them.

Taking the peculiar conditions of the individual countries into account, on the basis of reserve position alone, in the long run the following countries are judged to have higher discount rates: Saudi Arabia, Kuwait, United Arab Emirates, and Iraq. The following would have lower discount rates: Venezuela, Indonesia, Algeria, Nigeria, and Iran. Of all the measures, the effect of R/P ratios on discount rates is the most uncertain; the direction of effect is shown in column 6 of table 3-2.

*Country Discount Rates: Overall Evaluation.* The overall estimate of country discount rates is presented in column 7. Countries with an "L" are judged on the whole to have a lower preference for current income than for future income. Thus at a given price they would prefer to hold more reserves in the ground, rather than produce them immediately, in contrast to their preference if they had a high discount rate. This position is consistent with a high price–low

---

[13]Reserve-to-production ratios are very imprecise measures. Here the reserves are taken as proved and probable, not inventoried reserves alone. Possible new discoveries in unknown producing regions are not included.

output policy. The opposite policy would be attractive for a country designated with an "H."

This analysis indicates that Indonesia, Iran, and Nigeria can be expected to prefer a high output policy if that would increase current income, even if a somewhat lower price per barrel results. The United Arab Emirates, Kuwait, Qatar, and Saudi Arabia fall unambiguously into the low discount rate category. Judgments are uncertain for the other nations. Iraq and Algeria are relatively unknown, but their broader economies and relative sophistication would likely tilt them toward the high discount rate group. Venezuela is another question mark. While on the "social" scales it tends to rank with the low discount rate countries, its more balanced economy, its ability to use current oil revenues to "take off," and its petroleum reserve position (which makes it less concerned for possible spoilage of the market) tend to put it into the high discount rate group. In the short run, deliverability constraints may make Venezuela favor a higher price. That policy can in fact increase its current revenues, given its situation, and hence it is not inconsistent with the analysis here. Great uncertainty exists as to the position Libya may take. On the basis of factors which are subject to analysis, and because of its political situation, it most likely belongs with the high price–low output group.[14]

## Country Postures in Relation to the OPEC Price

The preceding analysis divides the OPEC countries into high and low discount rate groups. It thereby suggests which countries would tend to be satisfied with (or powerless to affect) an OPEC

---

[14]These conclusions appear to contradict the recent stated positions of some OPEC countries. The apparent conflict arises from the fact that OPEC spokesmen are stating their position at a point of time, while this analysis is long run. The world oil market is at present in disequilibrium. Current prices may be above the long-run cartel price. Some countries, nevertheless, are producing near their short-run capacity. Iran, in calling for higher prices, seeks to increase its current revenue, a position consistent with the view of its long-run goals stated above. In a period of induced oil shortage, higher prices, not lower prices, are consistent with Iran's high discount rate. Saudi Arabia, on the other hand, is calling for lower prices than those now existing. As the nation which will ultimately bear much of the responsibility for production cutbacks, and as the nation with the largest reserves, it fears that extremely high short-run prices will lead to long-run excess production. Ultimately, then, lower short-run prices will make higher long-run prices possible, without excessive future sacrifice on the part of Saudia Arabia. Again, in the context of 1974, Saudi Arabia is acting in a manner consistent with that predicted above.

price–output policy based on the U.S. self-sufficiency price. On economic grounds, the United Arab Emirates, Kuwait, Qatar, and Saudi Arabia would prefer higher prices and lower outputs than those consistent with the OPEC policy that will likely result. Unfortunately for these nations, they have no simple mechanism to enforce their demands upon the other countries. If they restrict output on their own, the long-run result would not be a significant change in the world price, but greater sales by competitors. The restricting countries would bear all the cost of reduced output and gain only a small share of the increased price. Perforce, the low discount rate countries must be satisfied by their market shares and even be prepared slowly to give them up to sustain the cartel that is so valuable to them.

Iran, Indonesia, Nigeria and possibly Algeria, Iraq, and Venezuela—at least in the long run—would favor an OPEC policy of lower prices and higher outputs if that were necessary to increase current revenues. In answer to the question posed earlier, these are the countries whose actions are most likely to upset the cartel for *economic* reasons. In their search for greater present revenues these countries can be expected to press continually to increase their own output to its physical limits.[15] The larger output they place on the market will bring temporary reductions in prices until nations with lower current income preferences reduce output.

The very nature of a cartel of sovereign countries with different discount rate preferences leads to cartel policy prejudiced toward higher production and lower prices than would be optimal for joint revenue maximization. This follows because those with higher preferences for current revenue have the ability to benefit themselves (though harming others) by unilateral action increasing production, while those with the opposite preferences cannot unilaterally restrict production without benefiting others and hurting themselves. The nonsymmetry of power to influence the cartel outcome leads to periodic declines in the cartel price. These declines set in motion self-correcting forces which, under most conditions, will reverse the decline and reestablish the original cartel price. Further analysis of

---

[15]The converse is, of course, also true. If at some point in time the country in question cannot supply its historic market or its "share" of the total OPEC market at current prices, it will press for a higher OPEC price. This appears to be the situation Venezuela faced in 1974. Venezuelan behavior in this disequilibrium situation does not alter the fact that Venezuela is prone to prefer current over future revenue, as compared to low discount rate countries.

the present value of reserves in the ground is needed to estimate the lower limit to which the cartel price can fall in response to such pressures without the cartel itself collapsing.

## THE LOWER BOUND OF THE CARTEL PRICE

The lower bound of the cartel price is that price which those countries favoring a high price–low output policy would act to maintain. Consequently, it is the "trigger" price below which the price cannot be pushed and the cartel survive. As the world price approached that lower bound, the affected countries would either support the price by limiting output or else retaliate against the price cutters, destroying the cartel. This trigger price for each country is found where the price netted back to the producer is equal to the present value of the oil in the ground.

### Determining the Country Trigger Price

The present value of expected future revenue depends upon the expected price, the discount rate chosen, and the period over which future revenues must be discounted. For reasons suggested above, the cartel price that will be expected by oil exporters can be taken to be $10 per barrel, netted back to the producing country. Appropriate discount rates for each country are suggested in column 8 of table 3-2. A few points should be noted with regard to the precise discount rates reported there. First, the important policy implications flow from the relative levels of discount rates, not from the precise correctness of the rates chosen. Consequently, an incorrect rate does not invalidate the analysis; it only alters the reservation price. Obviously the rank orderings are more meaningful than the absolute rates selected. Second, three main groupings of countries are identified: the high discount rate countries of Indonesia, Nigeria, and Iran; the low discount rate countries of the United Arab Emirates, Qatar, Kuwait, and Saudi Arabia; and finally, those that fall somewhere in between. Intensities within those groupings are indicated where differentials appear to warrant. Finally, these rates were selected to reflect long-term discount rates.

The selected discount rates must be applied over a specific period to find the present discounted value of the future revenue from a barrel of oil *not* produced today. Selection of the appropriate discount period presents difficulties. The choice made here is to discount the future income as though revenue from a barrel not produced today is received 15 years hence instead. Certainly none of the oil-producing provinces involved will be exhausted during this 15-year period, but physical exhaustion is not the sole or even the major factor in selecting a discount period. Political, economic, and technological elements are also important. Oil reserves may be made less valuable by development of other energy sources. Unforeseen cataclysmic events may make these—and all other current resources—obsolete. Destructive war may intervene. The ruling politico-economic structure may crumble. On a different level, the normal production curve of an oil-producing province peaks well before midway in the period over which production takes place. If production is bunched in the near end of a much longer producing period, a time horizon of 15 years might still be appropriate. Other time periods could have been chosen, and those who wish to do so may experiment with them.

Another and perhaps more productive approach would be to reason that a barrel of oil produced today lowers the *stream* of output over the future—it does not simply replace a barrel produced at some future *point* in time. An economic life of 30 years would be perhaps appropriate for this purpose, though again 30 years is probably too short to see the complete exhaustion of oil reserves. At low discount rates there is not much difference between the results using this procedure and the results using the 15-year period. The operative factors at the 2 percent discount levels are 0.7430 for the 15-year period compared with 0.7465 when the stream of output is spread over 30 years. At the 12 percent discount rate the difference is much larger: 0.1827 compared with 0.2685.

Table 3-3 contains the information necessary to calculate the lower bound or trigger price for each OPEC country (column 5). Column 1 contains the country f.o.b. netback for representative crude, taken from column 6 of table 3-1. The discount rate shown in column 2 is from column 8 of table 3-2. The netback price discounted over 15 years at the stated rate yields column 3, the present revenue-equivalent of a barrel of oil in the ground, on the assumption that in the future it can be sold at $10 c.i.f. the United States. The difference between the *actual* price the country gets for its oil (the f.o.b. price) and the oil's present revenue-equivalent measures the per barrel return to the OPEC member from maintaining the c.i.f. U.S. price at $10—it is the contribution the cartel makes to

TABLE 3-3.   Present Revenue Equivalent, Cartel Rent,
and Lower Bound Cartel Price for OPEC Countries

| OPEC Country | Long-run cartel f.o.b. price[a] (1) | Discount rate[b] (percent) (2) | Present revenue equivalent[c] (3) | Cartel rent (Col. 1 −3)[d] (4) | Lower bound c.i.f. U.S. price ($10—Col. 4) (5) |
|---|---|---|---|---|---|
| Algeria | $10.28 | 7 | $3.73 | $6.55 | $3.45 |
| Indonesia | 9.43 | 12 | 1.72 | 7.71 | 2.29 |
| Iran | 7.77 | 10 | 1.86 | 5.91 | 4.09 |
| Iraq | 7.57 | 7 | 2.74 | 4.77 | 5.23 |
| Kuwait | 7.01 | 3 | 4.50 | 2.51 | 7.49 |
| Libya | 9.91 | 7 | 3.59 | 6.82 | 3.68 |
| Nigeria | 9.58 | 10 | 2.29 | 7.29 | 2.71 |
| Qatar | 8.24 | 2 | 6.12 | 2.12 | 7.88 |
| Saudi Arabia | 7.45 | 3 | 4.78 | 2.67 | 7.33 |
| United Arab Emirates | 8.66 | 2 | 6.43 | 2.23 | 7.77 |
| Venezuela | 8.76 | 7 | 3.18 | 5.58 | 4.42 |

[a]From table 3-1, column 6.
[b]From table 3-2, column 8.
[c]Present value determined by applying discount rate for 15-year period.
[d]Representative crude adjusted to 34° gravity and 0.5 percent sulfur as explained in table 3-1.

country revenue. The final column represents the c.i.f. U.S. price at which, for each country, the cartel contribution is eliminated. This price is found by subtracting column 5 from the $10 cartel price. Note that the c.i.f. U.S. price is set by world supply–demand conditions and is the same at any point in time for all oil exporters.

## Application of the Model

An example of temporary cartel disruption and restoration will help in explaining the importance of the trigger price in setting a lower bound for the cartel price. Suppose that a country seeking additional current revenues has excess capacity and places it on the market on the assumption that its extra sales will have little effect on the overall world price. In other words, cartel discipline is insufficient to coerce each member, or else a member thinks it can violate cartel production plans without being discovered. The price-shaving country immediately obtains a greater share of the OPEC market. Instability in price results as other countries, to maintain their sales, retaliate by cutting price. If the competitive price cutting continued, Indonesia would be willing, in this example, to cut its price to as low as $2.29.[16] That is, even if Indonesia thinks that by waiting it can get $10 for its oil 15 years in the future, its high preference for current income will cause it to prefer to sell another barrel at any price above $2.29 c.i.f. United States rather than wait for that expected $10 price. This does *not* imply that Indonesia would prefer a $2.29 price to a higher price. Neither does it imply that Indonesia's revenues are higher at the lower price. A decline of the cartel price to $2.29 would not, of course, occur because forces would be set in motion to restrict the decline.

While Indonesia's reservation price is $2.29, other countries are in a far different position. The United Arab Emirates, for example, would be indifferent between a U.S. c.i.f. price of $7.77 and a future price of $10 according to the assumptions displayed in table 3-3. Hence, again assuming that it has faith in the eventual return of the $10 long-run price, the United Arab Emirates would prefer to cease production rather than to sell at a price below $7.77. It certainly would prefer to give up markets in the short run to Indonesia than to sell at "distress prices." This analysis, then, leads to the conclusion that those countries that have high cartel rents will increase their market share, in contrast to those that have low cartel rents. Countries with high reservation prices can be forced to "move over"; their oil is worth more to them in the ground than are the revenues from its sale at a price which is attractive to other countries. This process of changing market shares would continue until the countries which had a preference for a higher current revenue were satisfied, or until they were convinced that action to increase sales was self-defeating. The latter situation results when low discount rate countries retaliate against the original price cutters by being willing to sell at a lower price. This willingness results either from upward changes in their discount rate because of lower total revenues, or else from a policy decision to take short-term losses in order to discipline other members of the producing group.

## The Range of Cartel Prices

Starting down the list of those countries with the highest reserva-

---

[16]Remember that here we are comparing current revenue to the present value of future revenue. At no point do we consider the costs of production or transactions costs.

tion price, one can imagine both Qatar and the United Arab Emirates choosing to cease production temporarily or to reduce output to a minimum rather than accept a "too low" price. It is difficult to believe that a country with the petroleum market power and the pride of Kuwait would acquiesce in such price declines. It is impossible to believe that Saudi Arabia would allow its production to be pushed out of the market or allow its oil to be sold at a price below that oil's future revenue-equivalent to the Saudis. Before the U.S. c.i.f. price fell below the Saudi Arabian trigger price, either market shares would be adjusted to the satisfaction of all (and the $10 price would be reestablished), or else Saudi Arabia would discipline those who sought greater market shares by breaking the cartel. The enormous bargaining power of Saudi Arabia would thus constrain short-run fluctuations in the cartel price c.i.f. the United States between $10 and about $7.50 (see table 3-3). Given the assumptions made here, these are the prices U.S. producers can expect to compete with, provided the cartel succeeds.

It should be noted that the $7.50 lower limit (1973 prices) of the cartel price is very near the $7.88 price used in chapter 2 as a basis for the analysis of the U.S. energy market in 1985. These two prices were arrived at independently. The lower bound price is also approximately the average price for U.S.-produced crude in June 1974. Another similarity holds. The current price for oil produced under unregulated conditions is approximately $10 per barrel, or approximately the expected long-run self-sufficiency price. While the similarity of these prices proves nothing, it does reinforce the view that the assumptions made about the market in 1985 are not unreasonable.

It should be emphasized that a prediction of price variations between $7.50 and $10 depends only on each OPEC country seeking its own best interest. It requires no prearranged collective behavior to reach a superordinate goal. It requires no enforcement mechanism by an OPEC executive. In this model, so long as cartel members expect the eventual return to the $10 price, an automatic set of reactions constrains price movements. Put another way, the U.S. target price set by U.S. policy will determine the upper limit of the cartel price and affect the range of its fluctuations. What happens, however, if some outside force disrupts long-run cartel price expectations? The cartel would collapse irretrievably, and the world would be thrust into a competitive energy market. The conditions that might lead to cartel destruction are analyzed in the next section.

## DESTRUCTION OF THE OPEC CARTEL: PRECONDITIONS AND POSSIBLE CAUSES

Breakdown of the OPEC cartel depends upon the perception by the member countries that they can increase their sales if they are willing to shave their prices. M. A. Adelman has persuasively argued that oil-producing countries correctly realized that they did not have such power as long as they did not control the marketing of their own oil.[17] As noted in the first section of this chapter, until recently, international oil companies set production plans consistent with the quantity of oil that could be marketed at given prices. Within this framework, an increase in OPEC taxes left each international company in the same relative position. The tax (or royalty) was perceived by the companies as a cost. Any increase was passed on to consumers. The companies' margins, while substantial, did not offer sufficient latitude for severe price competition; the tax-paid cost provided a floor for price cutting in any case. The oil companies were complaisant tax collectors for the host countries.

When countries market their own petroleum, no such externally imposed lower limit on prices exists. Countries recognize that increases in revenues can accrue from selling more, even if at a price slightly below that which they have been receiving. The latitude for price cutting ranges from the existing f.o.b. cartel price down to production costs plus long-run scarcity value. When countries market their own oil, the difference between these two prices is cartel rent subject to division between consumers and producers. It is no longer a cost which each seller must pass forward if it is to stay in business. When countries rather than companies control the offer price for crude, and when countries set their own production plans, the world price can crumble. Expectations about price are more

---

[17]M. A. Adelman has written a number of works where this concept is fully explored. The most readable of these works is the article, "Is the Oil Shortage Real? Oil Companies as OPEC Tax Collectors," *Foreign Policy,* Winter, 1972–1973, pp. 69–107. The most complete presentation of the argument is found in *The World Petroleum Market.* A distinctly contrary viewpoint is expressed by James E. Akins in "The Oil Crisis: This Time the Wolf Is Here," *Foreign Affairs,* April 1973, pp. 462–490.

fragile, and collapse of the cartel comes when expectations about the long-run cartel price are fractured.

Control over crude oil marketing is in the process of passing to OPEC countries. Host countries obtain oil that can be marketed for their own account from three institutional arrangements: taking royalty crude in kind, operating national oil companies, and participating in joint ventures with foreign oil companies. The royalty option has existed throughout the history of oil production. Until recently, however, the agreed-upon company payment for royalty crude has been above the market price, and countries took little oil for their own account. National oil companies have always existed, but until recently most oil produced by them was also marketed by foreign firms. Participation with foreign companies is the most recent and most important way that countries have increased their control over production and marketing decisions. Participation agreements typically allow the host country to sell its share of the oil back to the operating firm, or else to market the oil itself.

Through these arrangements OPEC countries now control the marketing of a sizable share of the crude production flowing into international trade. That share is growing and has reached 100 percent in some countries. As a result, the countries now also control the margin between production costs and price. Consequently, oil companies have an incentive to "shop" among countries for lower prices which they can use competitively to increase their sales to ultimate consumers. Hence, the interests of international oil marketers and the producing countries no longer coincide. Whether some OPEC member countries will succumb to the temptation of shaving prices to increase sales depends upon whether initiating conditions are present. The previous section describing differences in member country preferences for current income suggests that they do. Whether price shaving results in mere revision of market shares, or whether it instead results in the destruction of the cartel depends on the dynamics of cartel behavior. Those dynamics are explored in the following sections where three of the possible ways in which the cartel may be destroyed are considered.

### Miscalculations of Bargaining Power

The bargaining power of the member country determines its influence in OPEC decisions. Bargaining power primarily affects a country's share of the total package of benefits from OPEC membership. Price is fixed in this model of OPEC's behavior, and thus market share becomes the primary matter at issue. The obdurate adversary posturings in secret council, the inflammatory public statements, the tentative moves to widen market share by shaving cartel price to make marginal sales—these represent moves to gain further advantage, though any advantage gained is lost to another. These conflicts, however intense, are generally held within limits. No country wishes to go so far in expanding its benefits as to bring about the kind of retaliatory behavior that might destroy the cartel. The amount of such activity that will be tolerated from each country depends on its bargaining power.

So long as the estimation of each country's bargaining power is consistent within the cartel—that is, so long as the country's and its fellow member's estimates are similar—stability is likely to be maintained. No weak country will move far enough to provoke retaliation; no strong country will demand so much as to create a suicidal impulse on the part of the weak. The moves of each country will be appropriately interpreted. A minor transgression of cartel rules will be recognized as a signal that the country believes it deserves a greater share of the cartel market. The response will be appropriate: resistance if others do not agree; a bigger share of the market if they do. There will be no concerted effort to discipline the errant member if others perceive he was acting within the general expectations of cartel behavior. The response to an action that is *outside* such expected behavior would be quite different. Each country then would expect that the action was designed to alter the basic world oil market structure, and the cartel might dissolve in bitter rivalry. The point is, so long as the power and posture of each country conform to its actual position and to its own self-image, mistakes about the meaning of actions are unlikely; the essential stability of the cartel can be reasserted even if it breaks down temporarily in internecine skirmishes.

The essence of predicting whether the OPEC cartel will survive is to determine whether the actual bargaining power of the countries coincides with their benefits from OPEC membership. To investigate this issue, the distribution of the benefits of OPEC membership must be determined. Then, it is necessary to ascertain whether that distribution is consistent with the bargaining power of the different countries.

Each country has a different capacity to produce, and hence its share of the output must be considered in relative terms. The most

obvious criterion for sharing output would be relative reserve size. If
bargaining power played no role in determining market share, the

TABLE 3-4.   Cartel and World Market Shares and Reserve Shares of
OPEC Countries 1970 and 1973[a]

| Country and year | Percent of OPEC output (1) | Percent of OPEC reserves (2) | Percent of world output (3) | Percent of world reserves (4) |
|---|---|---|---|---|
| Algeria | | | | |
| 1970 | 4.4 | 7.4 | 2.2 | 4.9 |
| 1973 | 3.5 | 1.9 | 1.9 | 1.2 |
| Indonesia | | | | |
| 1970 | 3.7 | 2.4 | 1.9 | 1.6 |
| 1973 | 4.4 | 2.7 | 2.4 | 1.7 |
| Iran | | | | |
| 1970 | 16.7 | 17.2 | 8.4 | 11.4 |
| 1973 | 20.1 | 15.3 | 10.9 | 9.6 |
| Iraq | | | | |
| 1970 | 6.8 | 7.9 | 3.3 | 5.2 |
| 1973 | 6.3 | 8.0 | 3.4 | 5.0 |
| Kuwait | | | | |
| 1970 | 12.3 | 16.5 | 6.0 | 11.0 |
| 1973 | 9.7 | 16.3 | 5.2 | 10.2 |
| Libya | | | | |
| 1970 | 15.1 | 7.2 | 6.8 | 4.8 |
| 1973 | 7.1 | 6.5 | 3.8 | 4.1 |
| Nigeria | | | | |
| 1970 | 4.5 | 2.2 | 2.3 | 1.5 |
| 1973 | 6.7 | 5.1 | 3.6 | 3.2 |
| Qatar | | | | |
| 1970 | 1.6 | 1.1 | 0.7 | 0.7 |
| 1973 | 1.9 | 1.7 | 1.0 | 1.0 |
| Saudi Arabia | | | | |
| 1970 | 15.4 | 31.6 | 7.7 | 21.0 |
| 1973 | 24.8 | 33.6 | 13.4 | 21.0 |
| United Arab Emirates | | | | |
| 1970 | 2.9 | 2.9 | 1.4 | 1.9 |
| 1973 | 4.3 | 5.5 | 2.3 | 3.4 |
| Venezuela | | | | |
| 1970 | 16.5 | 3.4 | 8.2 | 2.3 |
| 1973 | 11.3 | 3.6 | 6.1 | 2.2 |

Source: "World Wide Report," Oil and Gas Journal, December 28, 1970, pp. 92–93;
December 31, 1973, pp. 86–87.
[a]Totals may not equal 100 percent because of rounding. It should be noted that the
data on which this table is based are approximate. Reserves are not defined carefully,
and different countries are in different stages of exploitation maturity. The orders of
magnitude should, however, be representative.

hypothesis would be that a country's output would represent the same share of OPEC output as its reserves do of OPEC reserves, adjusted for exploitation maturity. These ratios do differ, however, as even the unadjusted data in table 3-4 demonstrate. Bargaining power may explain these differences. They may also be explained by different preferences as to rates of output.

If reserve-to-production ratios differ in a manner consistent with country bargaining power or output preferences, there is unlikely to be a disruption of the OPEC cartel. If those variations are not consistent, then disruption may occur. Country preferences as to price–output policies were considered earlier. An analysis of relative bargaining strength within OPEC is called for here.

*Determinants of Country Bargaining Power.* Bargaining strength within OPEC, as in other cartels, has two elements: bargaining power is inversely related to the harm to the subject country from cartel destruction; bargaining power is directly related to the ability of a country to destroy the cartel. The first element is a measure of vulnerability; the second is a measure of the power to sanction. Bargaining power arises both from the real position of a country and from other parties' perception of that position. The ability to threaten or to bluff is thus important. To be able to threaten successfully, a member must convince others that it is willing and able to take an action which will harm them, even though the action causes harm to itself. A successful threat never requires its performance. To bluff successfully is to persuade others that the member will withstand the threat and force the others to carry it through. A successful bluff precludes others from offering the threat. Bargaining strength, then, lies with those who are not vulnerable, those who have great power to do harm, and those who are seen to be willing and able to accept and return punishment.

A number of characteristics that affect country bargaining power can be identified. The factors of importance include production cost, relative size of reserves, importance of oil revenues, foreign exchange imports coverage, government stability, size of output, and security from external interference.

*Production cost.* Hypothesis: the higher the marginal production cost at the current rate of output, the lower the bargaining power. High production cost means that underselling as a retaliatory tactic is foreclosed to the member. Production cost among countries is so far below current prices however, that it is unlikely that any of the cartel participants anticipate prices falling to levels where production cost is important. To the extent that production costs are an element in bargaining strength, Persian Gulf countries are in a privileged position. Their strength is noted in table 3-5 by the placement of "G" for great bargaining strength in column 1.

*Relative size of reserves.* Three hypotheses exist. Hypothesis 1: the larger the size of the reserves (beyond the size which merely supports current production levels), the greater the bargaining power of the OPEC country. Excessive reserves increase the potential threat that a country can offer to world oil price stability. If not given its way, it has the ability to "discipline" the other cartel members through increases in output levels. Saudi Arabia and Kuwait have by far the greatest bargaining power on this basis, and receive a "G" in column 2 of table 3-5.

Hypothesis 2: the smaller the size of excess reserves (those above the level necessary to maintain current output), the greater the bargaining power. A country with reserves scarcely sufficient to support its current level of output will not elicit retribution when it acts in its own interest. It can do little harm. Yet, disciplining such a country has great costs. Hence, the transgressions of countries with small reserves are likely to be overlooked. For this reason, such countries as Venezuela, Algeria, Nigeria, and Indonesia have great maneuverability, if not bargaining power **per se**.

Hypothesis 3: countries with reserves consistent with output have small bargaining power. They do not have the reserve position which would enable them to threaten. They have no defense against retaliation. Yet their transgressions are not overlooked.

*Importance of oil revenues.* Hypothesis: the greater the proportion of oil revenues to the gross national product, the lower the bargaining power. The more important the oil revenues, the more vulnerable a country is to threats of others and the less credible are its own threats to "bring down" the cartel. Conversely, a country with a more balanced economy can face the trauma of disruption or diminution of oil revenues more nearly with equanimity. It can be aggressive or stubborn in turn.

*Foreign exchange imports coverage.* Hypothesis: the greater the accumulated foreign exchange compared to import requirements, the greater the bargaining power. Freedom of action flows, and is perceived to flow, from the ability to do without oil earnings.

*Government stability.* Hypothesis: the more internally stable a government *appears* to be abroad, the smaller the bargaining power. Weak governments are perceived to be unable to compromise

TABLE 3-5. Summary of Bargaining Power Indicators, OPEC Countries[a]

| OPEC Country | Factors affecting bargaining power[b] | | | | | | | Conclusion[c] |
|---|---|---|---|---|---|---|---|---|
| | Production cost (1) | Reserve position (2) | Importance of oil revenues (3) | Foreign exchange imports coverage (4) | External Government stability (5) | Size of output (6) | control vulnerability (7) | (8) |
| Algeria | S | G | G | S | G | S | S | S |
| Indonesia | S | G | G | S | G | S | G | G |
| Iran | G | S | G | S | S | G | G | G |
| Iraq | G | G | S | G | G | — | S | S |
| Kuwait | G | S | S | G | S | G | S | G |
| Libya | S | G | G | S | G | — | S | S |
| Nigeria | S | S | S | S | G | — | G | G |
| Qatar | G | G | S | G | S | S | G | G |
| Saudi Arabia | G | G | S | G | S | G | G | G |
| United Arab Emirates | G | S | S | G | S | S | S | S |
| Venezuela | S | G | G | G | G | G | G | G |

[a]The bases for the conclusions shown in this table are found in the text.

[b]"G" means that the country ranked among those with great bargaining power with reference to the factor considered; "S" means that the country ranked among those with low bargaining power; an empty cell implies indeterminacy with reference to this characteristic.

[c]According to the meanings given in the footnote above, the country *as a whole* is thought to possess either great or small bargaining power within OPEC.

domestic interests for external relations because they are subject to overthrow. Paradoxically, then, recognized ability to foreswear retribution, to act with magnaminity, or to act in the long-run interest of OPEC is a bargaining weakness. On the other hand, because it is known that a weak government will choose to sacrifice the cartel rather than be overthrown, its opponents will avoid putting it in that position. Apparently unstable governments can, therefore, narrow their options (by taking public positions), reduce their apparent bargaining space, and *improve* their bargaining position. As contrasted to apparent instability, actual stability improves the nerve and widens options; hence it increases bargaining power. The optimal condition is to be strong but to appear weak.[18]

*Size of output.* Hypothesis: the greater the absolute size of output, the greater the bargaining power. Large producers have great influence through the price impact of proportionately small variations in their production. Potential output expansion is captured in the reserve-to-output measure, and hence current output alone is the criterion here. Net exports, if obtainable, would be a better measure. Internal consumption is insufficient to affect the ranking of most major OPEC producers, however, and thus output alone is a satisfactory indicator.

*Security from external interference.* Hypothesis: the less secure a country is from external interference, the smaller the bargaining power. Insecurity may arise when a nation is vulnerable to external force of arms. Alternatively, a nation may have ties to outside forces or to governments that reduce its freedom of action. A strategic position may exist which induces large power interference. On these grounds the Arab states are vulnerable to the winds of Pan-Arabism. The small Persian Gulf states must look to neighbors or to major powers for protection. Large powers affect the freedom of still other member states, especially where a perceived threat to internal or external security is meaningful enough to make some sacrifice of freedom worthwhile. In explanation of column 7, the United Arab Emirates, Kuwait, and Qatar are vulnerable as small Persian Gulf states; Algeria and perhaps Libya are susceptible to threats from Pan-Arabism; Iraq is influenced by a large power, the USSR. Iran is

---

[18]An opposite argument can be made. The country that appears strong can make threats and promises that will be taken as binding over time. Thus there are some gains from appearing strong. On balance, however, to appear unstable in a world of popular uprising gives force to the negotiator who claims that he can surrender no more and maintain power.

relatively secure because of its own strength and because of the superpower standoff. Nigeria is secure because it has achieved a higher state of development than its near neighbors. Indonesia is secure because of its isolation. Saudi Arabia is strong because of its position, inherent strength, and control of the holy places of Islam.

*Summary of member country bargaining power.* The factors affecting bargaining power listed in table 3-5 are not of equal weight. An overall judgment of the relative strength of the OPEC countries within the councils of the cartel is indicated in column 8. Five countries have a preponderance of "G"s. Two of these, Indonesia and Nigeria, may be dismissed. They have maneuverability and thus can pursue their own goals reasonably freely, but they can do little to influence others. The giant of OPEC is preeminently Saudi Arabia; Venezuela is a distant second in influence. This conclusion is reinforced by other factors. The leadership of these countries has been good; each country is the leader of one of the two diverse groups within OPEC (Saudi Arabia among the rich, Arab, undeveloped countries and Venezuela among the poor, non-Arab, more balanced economies).

Positioning Iran is difficult; its standing is so important that perhaps it should be placed with Saudi Arabia and Venezuela. Its bargaining weaknesses are in part strengths in the special circumstances of that country. Moreover, its military power and effective leadership alone should perhaps place it in a higher position. It still appears, however, to rank well below Saudi Arabia, and probably behind Venezuela as well. The weakest of the OPEC countries are the United Arab Emirates and Qatar; they are almost clients of the superpowers within OPEC. The other countries fall more or less in the middle, with Kuwait, because of its low production cost, large reserves, and ample foreign exchange reserves at the upper end of the range.

*Cartel Destruction Resulting from Miscalculations of Bargaining Power.* A divergence between bargaining power and market share will give rise to potential disruption if a country with a low share and high bargaining power seeks to increase its share, but the other countries misjudge either its strength or its resolve. Tables 3-4 and 3-5 reveal that only Saudi Arabia qualifies as a country with high bargaining power and a share of output distinctly below its share of OPEC reserves. The other countries with less than a proportionate share of output are Kuwait, the United Arab Emirates, and Iraq. According to this analysis, these last three countries (with the possible exception of Kuwait) have low bargaining power and thus are in no position to enforce their demands on OPEC.

Consequently, Saudi Arabia is unique in its ability to increase its market share without serious threat of disruptive behavior on the part of other OPEC members. It will be seen to "deserve" more, and to have the power to obtain it. From table 3-2, however, note that Saudi Arabia (and Kuwait as well) prefers a high price–low output policy. If this estimate is correct, there may be good cause to believe that both Saudi Arabia and Kuwait are willing to foreswear an increase in their market share, and perhaps may even be willing to see their share slowly erode, in order to maintain the long-run high price–low output policy they prefer.

Attempted shifts in the market share of all other countries are likely to induce disruption. That disruption may or may not lead to the destruction of the cartel. The countries whose actions most likely would destroy the cartel are those which prefer high output in order to obtain more current revenues, those which already have more than their "share" of output, and those which are perceived to have low bargaining power or weak resolve. Referring again to tables 3-2, 3-4, and 3-5, it can be seen that *no* country fulfills all three of these criteria. Based on this analysis, it is unlikely that the usual intracartel jockeying for favored treatment will result in the kind of action and reaction that would disrupt OPEC. There is no inner dynamic that would predispose the OPEC cartel to internal breakdown from a divergence among goals, relative bargaining power, and reserve-to-production ratios for its member countries. Indeed, despite the mutterings and disagreements, thus far OPEC has been remarkably successful in that disagreements have been resolved without disruptive behavior, even though country positions diverge on many issues. On economic grounds alone, then, there is small possibility of a competitive market arising spontaneously from within OPEC because a continuation of accommodating behavior may be anticipated.

*Shifts in Country Conditions or Intercountry Relations*

Noneconomic or nonpetroleum factors can also destroy the cartel. In the analysis thus far it has been assumed that the member countries of OPEC are motivated largely, if not exclusively, by the desire to maximize the contribution of oil revenues to their other goals. It is known, of course, that factors other than these influence member country behavior. All such factors which might disrupt the cartel cannot be discussed here, but a few can be listed and one developed at more length. Among the possible disruptive developments that might be mentioned are a shift in member values, a failure of leadership, the disruption of non-OPEC relations leading to ties to some consuming country, and a change in the perception of the future. Intra-OPEC conflicts as a source of cartel disruption deserve extended discussion. Attention below is focused on such conflicts as they may arise in the Middle East and Africa.

The most obvious possible source of cartel disruption is political rivalry, especially in the Middle East where several forces now compete for ascendancy. Within the Arab camp there are radical socialist elements and traditional conservative ones. Animosity is great; it sometimes seems that hostility toward Israel provides the only cement that bonds the Arab peoples together. Disputes going back over 100 years separate the ruling families of Arab countries. Religious differences are great, even within normally tolerant and diverse Islam. Actual warfare, sometimes using client armies, has long existed in the hinterlands around the Persian Gulf. Rivalry over borders has been intense. Western observers often fail to recognize that nationhood in the Middle East is a twentieth-century phenomenon; borders were often imposed from without. Institutions are not settled and governmental legitimacy is not universally recognized. Conflict among the Arab OPEC countries thus is quite possible. Warfare using the oil weapon may well precede, or perhaps substitute for, military action. The cartel would disappear in a competitive scramble for markets should this occur.

Though Pan-Arabism is mostly a name for aggressive designs among the Arab inhabitants of this section of the world, the intra-Arab conflicts pale beside the Iranian–Arab rivalry. Iran's posture toward Israel has added to the inflammatory potential in the Middle East. Moreover, Persia is strong, expansionist, and proud—ever mindful of threats from the north, from its western borders, and from across the Gulf. A showdown with Saudi Arabia in which zones of influence are negotiated or fought out is almost inevitable. Should Iraq become more stable internally, Iranian accommodation with that potentially powerful country would also be required.

Every possibility exists, then, that OPEC interests might be sacrificed in the heat of these international pressures. Extremely shrewd leadership on the part of all the participants will be required to maintain cooperation in the potentially explosive periods ahead.

That leadership may be difficult to maintain during the next decade when death will likely bring a number of changes, even if political events do not. As the oil-producing countries' needs for additional revenue become less intense, geopolitical rather than economic goals may well come to appear more important. In this light, the confidence in OPEC survival that is based on analysis of economic issues cannot be matched by like confidence when noneconomic prospects are considered. Direct intra-OPEC conflict would, of course, disrupt the cooperation necessary for cartel survival. Once open conflict broke out, it is difficult to believe that OPEC could ever be reestablished at its current level of authority. A competitive market would result.

### Shifts in External Market Factors

The survival of the OPEC cartel depends on the stability of the market forces affecting the demand for and supply of OPEC petroleum. When those forces change rapidly, destruction of the cartel may follow. The cartel is especially vulnerable to a disruption of price expectations. As suggested above, the cartel coalesces around an expected price. If that price is altered, and if no satisfactory mechanism exists to readjust member claims at a new focal price, all cartel expectations may fall. The cartel structure will tumble after, and a competitive oil market would ensue.

The most likely exogenous factor which might disrupt the cartel would be a dramatic reduction in the demand for OPEC petroleum. This reduction could arise either from major oil discoveries outside the OPEC countries or else through a dramatic technological breakthrough which would lower the expected value of all energy sources. Heretofore OPEC has been successful in bringing new exporting countries into the cartel. This policy will fail if a new major producing region is opened within a major importing country market. The North Sea discoveries have already improved Britain's condition. Norway will soon become a net exporter. OPEC's expected market in Europe has already fallen by 6–8 MMb/day. A discovery of like magnitude in Japanese waters would cause a substantial prospective shift in the demand for OPEC oil. Such a discovery could by itself fracture OPEC expectations and lead to lower prices. Technological breakthroughs that shorten the expected time horizon for the onset of new energy sources would have the same effect. A development which conserved energy—a low-cost superconducting material, for example—would result in a direct demand reduction. Again, disruption could occur.

Massive new discoveries within OPEC (or in a country that joins OPEC) would also place strains upon the cartel. Current producers would perhaps be required to forgo current sales in addition to market growth. The adjustments of market shares would be difficult, especially where nations had adjusted to rising export earnings.

The cartel could not adjust to demand reductions or supply increases that totally swamped price expectations. A competitive price would inevitably result. The cartel might, however, be able to adjust to a small demand decline or a supply increase by one member either by restricting output or by letting the price erode gradually and in an orderly way. In either case, the cartel would need to select and adjust to a new price–output policy. Adjustments of this sort are difficult. During the delicate process of adjustment, intracartel negotiations might easily disintegrate into full-scale competition. If competition ever broke out, there would be no clear boundaries to limit it and a world competitive price would follow.

The analysis given here indicates that a competitive world market may arise from at least three developments: faulty evaluation of the bargaining power of a producing country, deterioration in intracartel relations, and changes in external market factors. Other disruptive possibilities exist. Whatever the source of disruption, the ability of the cartel to hold the price of petroleum above the competitive level would disappear. When contemplating possible U.S. policy, it is important to know what the long-run competitive price for petroleum delivered in the United States would be. An approach to determining that price is presented in the following section.

## THE LONG-RUN COMPETITIVE
## PRICE FOR PETROLEUM

The price of imported oil delivered to the United States is equal to the sum of production costs, payments to resource owners, transportation, and handling costs. The "competitive" long-run price, or the price if the cartel were to break down completely, would be one which covered the production cost of the marginal barrel, a payment to cover the long-run cost of transporting the marginal barrel to the United States, and the minimum return to individual resource owners necessary to induce them to sell their oil rather than hold it for future sale. Production and transportation costs are relatively well defined. The necessary payments to host countries are not. As analysis of the cartel price suggests, these payments are based in a complex way on the resource owner's expectations about future demand and supply conditions for energy. In estimating the level of such payments under competitive conditions, the Persian Gulf may be taken as the price-determining source of oil. That region will provide the major source of incremental supplies to the world for the foreseeable future.

## Production Costs

The first element of the competitive price to be estimated is production costs around the Persian Gulf. Production costs include exploration expenditures required for finding oil deposits to replace those being consumed, development expenditures required for wells and downstream facilities to the port of exit, and operating costs over the life of a reservoir. The earlier work of Adelman is still the most thorough available on oil production costs.[19] Adelman's basic approach was to estimate the investment required at the margin to find and develop the oil reserves required to offset the reserves produced, and then to amortize this investment with an appropriate discount rate to determine a price that just compensated the operator for his investment. Annual operating costs were then added to investment costs to obtain total production costs per barrel.

Adelman's cost estimates for 1985 were based on a set of assumptions designed to overstate the marginal cost of oil produced outside of North America.[20] He assumed that natural gas and nuclear energy would not be available for energy consumption, non-U.S. coal would be exhausted, no new oil discoveries would be made before 1985, no new technological developments would occur, and the decline rate in the Persian Gulf would increase. Persian Gulf output was assumed to grow from 13.65 MMb/day in 1970 by 11 percent per year to 24 MMb/day in 1985, reducing the reserve-to-production ratio in 1985 to 16:1. Adelman seriously underestimated the rate of

---

[19]M. A. Adelman, *The World Petroleum Market,* and "Oil Production Costs in Four Areas," Proceedings of the Council of Economics of the American Institute of Mining, Metallurgical and Petroleum Engineers, Inc., Annual Meeting, New York, New York, February 28–March 2, 1966.

[20]*The World Petroleum Market,* chapter 2.

growth of Persian Gulf production; it had reached 20 MMb/day by 1973. The current reserve-to-production ratio is about 50:1, however, so his production cost estimate may nevertheless be sound. Independently estimated costs of Saudi Arabian crude tended to decline or remain constant into 1974, even in current dollar terms, in spite of rapidly expanded output.[21] Adelman concludes that the 1985 marginal cost of producing oil in the Persian Gulf would be no more than 20 cents per barrel in 1968 prices. His estimates are accepted here.

*Transportation Costs*

The transportation cost in the long-run competitive price is determined by the long-term charter rate for tankers from the Persian Gulf. The tanker market is competitive, so trends in market prices reflect long-run supply prices.[22] The American Committee for Flags of Necessity has predicted that by 1980 the Persian Gulf–East Coast long-term charter rate will be equivalent to $1.37 per barrel (in 1973 prices) on a foreign-flag, 70,000-deadweight ton tanker (the corresponding cost for a U.S. flag tanker is $2.07 per barrel).[23] Future transport costs could be lower if American ports were equipped to take advantage of the economies of scale of the very large crude carriers. A policy of self-sufficiency would forestall the timely development of required port facilities, however, so the indicated transport savings cannot be assumed. It seems safe to conclude that the equilibrium tanker rates could decline in the future below the $1.37 per barrel cost suggested above. Just how far below is speculative, and depends in part on future U.S. import policy.

*Payments to Resource Owners*

A competitive market would include in the price of oil a payment to Persian Gulf countries sufficient to induce them to sell their oil reserves now rather than hold them for future sale. The reservation price of the oil-exporting countries will include different institutional components of which only one, an ordinary business tax, need

---

[21]*Middle East Economic Survey,* December 28, 1973, p. 6.

[22]*The World Petroleum Market,* chapter 4.

[23]American Committee for Flags of Necessity, *Tankers and Oil . . . The 70's,* July 1973, p. 2.

be considered explicitly here. Some part of the revenue received by exporting governments from the sale of oil is in lieu of normal taxes paid by any entity to support government operations. It is not possible to generalize about what level this tax would be, and hence it can only be noted that not all of the revenues to host countries would cease even if the value of oil in the ground were decreased to zero.

Owners will be motivated to sell their reserves whenever the present value of the expected future return is less than the price. Left in the ground, oil can produce a return for its owner by appreciating in value. To be in equilibrium, resource owners must expect the royalty to increase at a rate equal to the rate of interest. If royalties are expected to increase at a faster rate, resource owners will demand higher royalties today or else slow extraction; at a lower rate they will speed extraction.

The rate of increase in royalties depends upon the prospective scarcity of the resource; scarcity is a function of the rate of technological progress toward substitutes. The process just described is the market's way of conserving a scarce resource until an alternative becomes available, and then of disposing of the remaining reserves quickly as their use value falls toward zero.

William D. Nordhaus estimated the scarcity value or royalty for Middle Eastern crude to be approximately $0.18 per barrel in 1970, rising only to $0.44 in 1980.[24] The Nordhaus estimates were based in part on generally conservative assumptions, but they also were premised on a level of certainty in the development of alternative energy sources that participants in the market are not likely to share. Moreover, the reservation price for producers in a competitive market depends on the collective expectation of what the oil price in the future will be. These expectations are in part self-validating; surely even unjustified expectations can exist for a long period, and hence the Nordhaus estimates are not dispositive, even if correct.

On the basis outlined here, it seems unreasonable to expect exporting countries to be willing to transfer their oil reserves for anything less than they received before the recent escalations. In other words, they may be expected to perceive the competitive payment as the amount they received *before* OPEC exercised its cartel power. Royalties below that amount would be rejected as exploitative; major production would be stopped. Note, too, that royalty payments

---

[24]William D. Nordhaus, "The Allocation of Energy Resources," *Brookings Papers on Economic Activity,* 3, 1973, pp. 529–570.

include an undetermined amount in lieu of ordinary business taxes. For Persian Gulf oil, then, a reasonable long-term transfer price in a noncartelized market can be expected to be approximately $1 per barrel. While no certainty can be claimed for that figure, it is unlikely to be significantly wrong.

### The Delivered Price of Oil:
### The Competitive Case

The long-run noncartel price of oil delivered in the United States is the sum of the estimated costs described above plus a margin to cover handling costs and any tariffs imposed. The sum of the costs plus a reasonable handling charge would come to approximately $3/bbl.

The $3/bbl estimate is a long-run price that would be subject to great fluctuations. Energy production is capital intensive, and the capital involved is long-lived. As a result, a significant time lag is required for the quantity of energy supplied to adjust to a shift in demand. Short-run supply elasticity is very low, and hence exogenous shifts in demand are primarily accommodated through price changes. Similarly, because of the capital-intensive nature of complements to energy consumption and the paucity of short-run substitutes, the demand for energy is very price inelastic in the short run as well. Hence an exogenous shift in supply will result primarily in a price, not a quantity, response. Under competitive conditions, then, in a changing world we would expect fluctuating energy prices around the $3 level.

The recent experience with petroleum prices is instructive. The rapid increases in world prices following the simultaneous exogenous increase in demand and the small supply reduction by Arab countries emphasize the volatility of this market. A competitive world oil market would face similar fluctuations in underlying prices, but the operation of hedging markets and the development of commercial traders and storage concerns would keep variations within lower bounds than those observed recently.[25]

---

[25]Parenthetically, the past ability of the international companies to control production and to develop markets in an orderly way foreclosed the creation of firms engaged in these intermediary economic functions. Now that supply insecurity is a recognized reality, one could anticipate commercial ventures which seek profit by stabilizing the market.

The competitive market described here can occur only if the OPEC cartel is disrupted. Similarly, of course, it would likely be foreclosed by development of a consumer's cartel or by a drive for self-sufficiency on the part of the United States. A discussion of the possible outcomes with reference to the price and availability of oil follows.

## Judgments on the Future World Price

The price of oil delivered to the U.S. will be approximately $3 per barrel in the competitive case, according to the models and the assumptions developed in this chapter. If the OPEC cartel remains secure, the price will range from a low of $7.50 during periods of temporary cartel disarray to $10 (in 1973 prices) when the cartel is functioning smoothly.[26] The range of expected prices is obviously very large. The implications for U.S. energy policy are correspondingly diverse. If the United States could confidently expect a competitive world oil market free of interruptions, a *laissez-faire* policy would appear optimal—energy security would be cheap. Similarly, a *certain* $10 price would not require government action because, again, energy security would be obtained. In this circumstance security would arise from action of the private market leading to domestic self-sufficiency. Neither of these certainties exists.

Instead, U.S. policy must deal with uncertainty in the price of oil. From the analysis here it appears that the wisest course would be to premise U.S. policy on an expected delivered price of foreign oil of between $7.50 and $10 per barrel. This recommendation reflects the view that the OPEC cartel will remain intact. This view, in turn, is premised in part on the judgment that the United States will pursue somewhat autarchic energy goals. There will be almost perpetual jockeying for sales within OPEC, and hence the oil price will more often approach the lower bound of the cartel price than the upper. It seems unlikely, however, that this lower bound will be pierced and the cartel destroyed. The precondition for cartel destruction—country control over marketing—exists. Those countries that have the power to disrupt the cartel, however, do not have the incentive;

---

[26]These estimates are consistent with the "self-sufficiency" mindset in the United States. Chapter 6 develops a different outlook as to the future course of world oil prices, based on possible U.S. initiatives to encourage external investment by oil exporters.

those countries that have the incentive (the will to increase their output) do not have sufficient reserves to threaten the cartel. Periodic price wars will occur. They are unlikely to get out of hand. Certainly U.S. policy should not be based on the expectation that they will do so. Neither, of course, should it be based on the expectation that foreign oil will always be $10 or more per barrel.

The judgment here is that there is insufficient reason on economic grounds to expect a breakdown of the cartel. Disruption from other causes is not precluded. From now until 1985 the major danger to the cartel will come from noneconomic pressures within it. Factors other than oil economics can potentially inflame the oil-producing countries of the Middle East and Africa. There is, nevertheless, reason to believe that these governments will be able to accommodate themselves to the shocks which may come. The success of OPEC thus far, the coincidence of economic interests of the most powerful of the member countries, and the force of a common bond against what is perceived to be exploiting foreign interests may in combination be sufficient to hold the inevitable conflicts in check. The world has witnessed numerous examples of conflict among countries on one level while economic cooperation continued on another.

Shifts in external market factors are the least likely potential source of cartel disruption. Truly immense increases in energy supply could not reach the market much before 1985. The important question is whether by that time cartel expectations as to future oil prices could be altered by new discoveries or technological change. Given the relatively short time horizon posited in this study, a substantial change is unlikely. Small changes in expectations could be accommodated. For the period beyond 1985, different conditions hold. Exogenous factors can be so diverse and their influence so large as to make prediction that far in the future unproductive.

Much uncertainty exists as to the course oil prices will take during the period up to 1985. The OPEC countries have not yet faced up to the need for joint production control. Prices may well decline precipitously before the need for discipline within the cartel overcomes the reluctance of each to subordinate its goals to the interest of the collective. A short-term initial decline below the lower bound cartel price could be tolerated without disruption of the expectation that the cartel in the end would prevail. Collapse after a production programming system was established would be much more serious for the future of the cartel.

The conclusion is, then, that U.S. policy must prepare for existence of an OPEC cartel with the ability to interdict U.S. supplies in the short run. The cartel will usually express its strength by holding the world price of oil within a range of $7.50 to $10 a barrel. In calculating the cost of energy security, it is these upper-bound and lower-bound cartel prices which are most relevant for the term through 1985.

# 4

# Tariffs, Quotas, and Self-sufficiency

Chapter 2 presented an estimated relationship between the domestic price of oil and the quantity of imports, based on the specified projections for domestic energy supply and demand. It was also noted that a given oil price-import level may be achieved with the use of either tariffs or quotas. However, the need for restriction depends on the price of imported oil. If the landed price remains above $10/bbl, there is no need to impose import controls because domestic supply and demand will adjust to eliminate imports. Standby controls might be necessary, but only to validate the expectations of domestic producers and consumers that the landed price would remain above $10/bbl. Similarly, if the landed price remained at $9/bbl, imports would fall to around 20 percent of domestic production, or about the amount of imports expected from the Western Hemisphere in 1985. If this volume of imports does not jeopardize economic security, then again no import controls would be required except perhaps on a standby basis. Thus, if the world price remains high enough, there is no need to impose import controls because the response of domestic producers and consumers to the price would adjust imports appropriately.

The point of reference for this chapter is to suppose that the world price is expected to fall to a level that encourages a volume of imports that is inconsistent with national security. Chapter 3 estimated that the landed price could fall to as low as $3/bbl if the world market became competitive. However, the analysis in chapter 3 suggests that the OPEC cartel will not dissolve before 1985. As a result, the landed price of imported oil will most likely run about $7.50/bbl or more.

The $7.50/bbl cartel price corresponds roughly to the assumed price used to obtain the initial projections in chapter 2. The import projections in the base case, therefore, imply no import controls at the cartel minimum price. The same projections imply import controls if the competitive world price should prevail. Imports would be limited to the projected supply–demand gap at the assumed price ($7.88/bbl).

The $3 minimum competitive price and the $7.50 minimum cartel price provide convenient reference points at which one can compare the costs and effects of tariffs and quotas. The protective effect and the costs of import controls depend on the difference between the domestic price maintained by controls and the import price that would prevail in the absence of controls. Given the level of domestic price to be maintained, the gap would be at a maximum when compared to the $3 competitive lower bound price. The maximum gap expected under cartel conditions is that based on an import price of $7.50/bbl. As mentioned before, the gap would approach zero (and so would the costs and protection of import controls) as the import price increased above $7.50/bbl. Thus, the $3 and $7.50 prices provide the basis for estimating the maximum costs of import controls under two extreme world market conditions: competition and monopoly.

## Economic Effects of Import Controls

The benefit of import controls lies, of course, in the presumed energy security gained through the reduction of imports. This is achieved by raising the price in the domestic market above the import price, which encourages additional domestic production and discourages domestic consumption. The relationship between prices and imports was estimated in chapter 2. The results are reproduced in table 4-1 for convenience.

Associated with the benefit of reducing imports are a variety of economic costs. The most obvious is the fact that consumers must pay more for petroleum products if the domestic price is higher than it would be with free trade. Not only do they pay more for what they purchase, but the higher price also means that they will buy less than they would at the free trade price. That is, consumers are paying more (the direct consumer costs) and they are getting less

TABLE 4-1. Estimated U.S. Price–Import Relationship in 1985

| Import reduction | Price ($/bbl) | Domestic production (MMb/day) | Domestic consumption (MMb/day) | Imports (MMb/day) | Imports as a percent of | |
|---|---|---|---|---|---|---|
| | | | | | Production | Consumption |
| Base case | | | | | | |
| Midpoint estimate | 7.88 | 14.6 | 22.2 | 7.6 | 52.1 | 34.2 |
| Upper estimate | 7.88 | 12.3 | 27.1 | 14.8 | 120.3 | 54.6 |
| Ban Arab imports | | | | | | |
| Midpoint estimate | 8.43 | 15.6 | 21.4 | 5.8 | 37.2 | 27.1 |
| Upper estimate | 10.29 | 16.1 | 23.0 | 6.9 | 42.8 | 30.0 |
| Ban non-Western Hemisphere imports | | | | | | |
| Midpoint estimate | 9.14 | 16.9 | 20.4 | 3.5 | 20.7 | 17.2 |
| Upper estimate | 11.32 | 17.7 | 21.2 | 3.5 | 19.8 | 16.5 |
| Ban all imports | | | | | | |
| Midpoint estimate | 10.24 | 19.0 | 19.0 | 0.0 | 0.0 | 0.0 |
| Upper estimate | 12.39 | 19.3 | 19.3 | 0.0 | 0.0 | 0.0 |

(the indirect consumer costs). An important question is where the additional consumer expenditures go: to whom and for what. Part of the increase must go to pay for higher-cost domestic production. The additional cost of developing domestic resources rather than importing cheaper foreign oil constitutes a resource cost to the American economy. Another part of the expenditure is transferred to producers for oil that would have been produced without the price increase. These are revenues over and above those necessary to cover costs and thus constitute a windfall gain because of import controls.[1] The revenues involved amount to a transfer of income from consumers as a group to producers as a group.

A final part of the increased petroleum bill for consumers is the difference between the price paid for imports allowed under controls and the price these imports are sold for in the domestic market. This differential is called a "scarcity rent" for imports, because it is an increase in the value of imports artificially created by import controls and not a payment necessary to attract imports. This difference would be the import duty if a tariff system were used, or the license fee if quotas were auctioned to importers. If quota licenses were not auctioned—they have never been so far in U.S. experience—then the revenues would accrue to either importers or exporters, depending on market conditions and how the licenses were allocated. If the importer had the opportunity to fill his quota from any exporting country, and if exporting countries bid competitively for the sale, then the importer would receive the scarcity rent of imports.[2] If an effective oil cartel were maintained, exporters might require the full domestic price (less delivery costs) from importers, in which case the exporter would receive the scarcity rents. Alternatively, if the import licenses were allocated according to country of origin, the exporting country would have the bargaining power and would tend to set the price to importers at the higher domestic level (less delivery costs) and capture all of the scarcity rents. Consequently, unless a tariff is used or quotas are auctioned, the scarcity value of imports is likely to accrue at least in part to oil-exporting countries.

The relationship among consumer costs, resource costs, income

---

[1]Schemes designed to control prices received for "new" and "old" oil attempt to distinguish between the differential production costs of existing production and new production. These controls may succeed for only a short term; eventually the two prices must converge.

[2]If there is extensive competition among importers, they may pass along part of the scarcity rents to final consumers in the form of lower product prices.

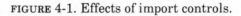

FIGURE 4-1. Effects of import controls.

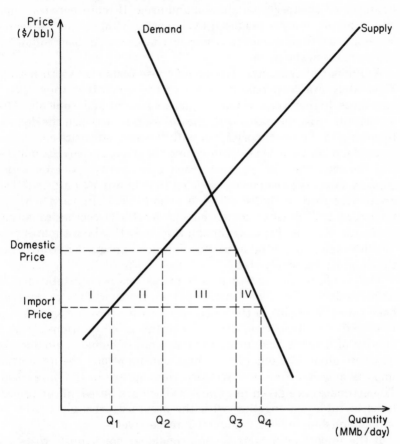

transfers from consumers to producers, and the scarcity rent of imports may be illustrated with the use of figure 4-1. The domestic price is drawn above the import price, the difference being due either to the tariff or to quota restriction. As a result of the price increase, domestic production rises from $Q_1$ to $Q_2$, domestic consumption falls from $Q_4$ to $Q_3$, and therefore imports decline from the difference between $Q_1$ and $Q_4$ to the difference between $Q_2$ and $Q_3$. The direct consumer cost described above is the price differential times the quantity consumed ($Q_3$), or the sum of areas I, II, and III, in the figure. Area I reflects the transfer of income from consumers as a group to producers as a group; area II reflects the resource costs of producing at home the increased quantity from $Q_1$ to $Q_2$ that could be imported cheaper from abroad; and area III represents the scarcity value of imports produced by import controls. Area IV represents the loss to consumers as their purchases are cut back following the increase in the price.

A number of characteristics about these costs are worth noting. First, they are "opportunity" costs in the sense that they reflect payments higher than would be necessary without controls. The opportunity cost increases with the *differential* between the domestic price and the import price, not with the domestic price alone. For example, if both the domestic price and the import price rise, but the gap between the two narrows, then opportunity costs have decreased. Also, the reader is reminded that in order to estimate the areas described in figure 4-1, one hypothetical situation without import controls must be compared with another hypothetical situation with controls. Tenuous though they may be, the estimates provide a range of potential outcomes that are important in conveying the magnitudes involved.

The range of alternative import prices has been established at $3/bbl and $7.50/bbl. The domestic price alternatives range from the base case ($7.88/bbl) to the price required for zero imports. As an intermediate alternative, we consider the price relationship at a volume of imports amounting to 20 percent of domestic production (or, equivalently, about 17 percent of consumption—the amount of imports expected from the Western Hemisphere in the base case). The estimates are given in table 4-2. They are based on the projections in table 4-1 together with the assumed price elasticities of supply and demand of 1.0 and $-0.5$, respectively.

If the import price were $3/bbl, consumer opportunity costs are estimated to range from $23 billion in the base case to $66 billion in the case of complete self-sufficiency. If the import price were $7.50/bbl, consumer opportunity costs are estimated to range from zero in the base case to $34 billion in the case of complete self-sufficiency. It is clear that the consumer price for self-sufficiency is very expensive in either case, but it doubles if world prices are reduced to a competitive level.[3] It is sometimes overlooked that if the United States wishes to pursue a policy of restricting oil imports,

---

[3]One need only recall the reaction to estimates of the opportunity cost of the Mandatory Oil Import Program (about $5 billion for consumers) in 1969 to appreciate the political sensitivity of these figures. *The Oil Import Question,* A Report by the

TABLE 4-2.  Opportunity Costs of Import Controls
Under Different World Prices and Domestic Conditions

($ billions per year)

| Domestic condition | World price | |
|---|---|---|
| | $3.00/bbl | $7.50/bbl |
| Base case ($7.88/bbl price) | | |
| Midpoint estimates | | |
| Direct consumer costs (area I + II + III) | 23.3 | 0.0[a] |
| Transfer costs (area I) | 10.6 | 0.0 |
| Resource costs (area II) | 4.7 | 0.0 |
| Scarcity rents (area III) | 8.0 | 0.0 |
| Upper estimates | | |
| Direct consumer costs (I + II + III) | 28.5 | 0.0 |
| Transfer costs (I) | 9.0 | 0.0 |
| Resource costs (II) | 4.0 | 0.0 |
| Scarcity rents (III) | 15.5 | 0.0 |
| Limiting imports to 20 percent domestic production | | |
| Midpoint estimate ($9.14/bbl price) | | |
| Direct consumer costs (I + II + III) | 40.4 | 6.8 |
| Transfer costs (I) | 23.3 | 0.4 |
| Resource costs (II) | 11.2 | 5.2 |
| Scarcity rents (III) | 6.9 | 1.2 |
| Upper estimate ($11.32/bbl price) | | |
| Direct consumer costs (I + II + III) | 56.4 | 21.3 |
| Transfer costs (I) | 29.8 | 2.7 |
| Resource costs (II) | 17.3 | 15.1 |
| Scarcity rents (III) | 9.3 | 3.5 |
| Ban all imports | | |
| Midpoint estimate ($10.24/bbl price) | | |
| Direct consumer costs (I + II + III) | 50.2 | 18.9 |
| Transfer costs (I) | 32.5 | 16.7 |
| Resource costs (II) | 17.7 | 2.2 |
| Scarcity rents (III) | 0.0 | 0.0 |
| Upper estimate ($12.39/bbl price) | | |
| Direct consumer costs (I + II + III) | 66.2 | 34.4 |
| Transfer costs (I) | 45.1 | 30.2 |
| Resource costs (II) | 21.1 | 4.2 |
| Scarcity rents (III) | 0.0 | 0.0 |

[a]The difference between the $7.88 and $7.50 estimates is not significant enough to calculate any opportunity costs. The other cost estimates could be rounded off in order to avoid giving the impression of undue precision, but this makes it difficult for the reader to duplicate the calculations.

Cabinet Task Force on Oil Imports Control (Washington: U.S. Government Printing Office, 1970).

even partially, that policy will be less expensive (and more palata-
ble) if the OPEC cartel succeeds in holding the price above the
competitive level.


## TARIFFS VERSUS QUOTAS

It was noted earlier that either tariffs or quotas could be used to
achieve the various combinations of domestic prices and import
levels discussed in the previous section.[4] Despite this fact, there are
a number of important differences between tariffs and quotas that
may establish a preference for one or the other as a policy instru-
ment. Tariffs and quotas differ in their (a) focus of control (price
versus quantity), (b) discriminatory properties, (c) administrative
flexibility, and (d) scarcity rents from imports. Preferences between
tariffs and quotas arise from these characteristics.


### Price versus Quantity Control

A quota system clearly establishes the maximum quantity of imports that will be allowed to enter a country.[5] Competing domestic
producers need not fear additional imports if domestic prices rise.
Consequently, this aspect of quotas makes them very appealing to
existing domestic producers. There is correspondingly great uncertainty about the effect of a given quota on the domestic price. Restricting imports requires the domestic price to rise enough to generate additional output and discourage consumption to make up for
the decline in imports. How much the price must rise is uncertain,
even if we assume that producers and consumers can react immediately to any price change.[6] The time lag required for changes in

---

[4]It is important to note that tariffs and quotas may be imposed in special ways that
will result in differences in the effects of the two control devices. For example, the
mandatory quota imposed in 1959 limited imports to a percentage of domestic production. It can be shown that the tariff equivalent of this approach would impose higher
costs to consumers. See George A. Hay, "Import Controls on Foreign Oil: Tariff or
Quota?" *American Economic Review*, vol. LXI, no. 4 (September 1971), pp. **689-691**.

[5]The quantity may actually be less than the maximum in the event that the foreign
price is so high that the demand for imports is less than the quota allocation.

[6]It will be recalled that the price–quantity estimates in table 4-1 assume enough
time has elapsed to allow for any adjustments.

output and consumption increases the price uncertainty.[7] Consequently, the allowed quantity of imports is often set at the current level in order to avoid the difficult question of the price effect. There is no immediate price effect, but imports cannot further replace domestic production or share in the growth of consumption. Of course, this approach is not very helpful if the current level of imports is judged too high for national security.

Tariffs, on the other hand, provide relatively greater certainty about the resulting domestic price, but less certainty about the quantity of imports. The new price is simply the old price plus the tariff duty. The tariff rate may be chosen with the intention of increasing the prevailing price to a specific target level. This approach is appealing for potential domestic production from heretofore uneconomic sources.[8] For production using new technology, the expected price is more important than the amount of oil that will be imported. In other words, the ability to enter the market in the first place is more important than the share of the market taken by imports.

The tariff may be adjusted to alter the quantity of imports desired. Similarly, the quota may be adjusted to alter the impact on the domestic price. Changes of this sort raise a question of relative administrative flexibility that will be addressed shortly.

As demand and supply conditions change over time, the distinction between tariffs and quotas becomes even more important. It may be useful to illustrate the differences with figure 4-2. Figure 4-2 reproduces figure 4-1, except that the demand curve is shifted outward to reflect an increase in demand in response to, for example, a growth of income or industrial production. Before an increase in demand, the effects of a tariff and a quota may be said to be identical. Areas I, II, III, and IV are the same in either case; only the uncertainties about the magnitude of these areas are different, as noted.

Now suppose demand shifts outward. With the tariff system, the domestic price remains unchanged at $P_2$ as long as the import price does not change. The increase in demand is accommodated by an

---

[7]There is also less urgency for producers to adjust to a price increase. Without the worry of foreign encroachment in domestic markets, there is one less source of competitive pressure to force producers to react to the change.

[8]For example, the target price may be set to yield an attractive rate of return on the production of synthetic crude oil and gas from coal or oil shale.

FIGURE 4-2. Comparing a tariff and a quota with an increase in demand.

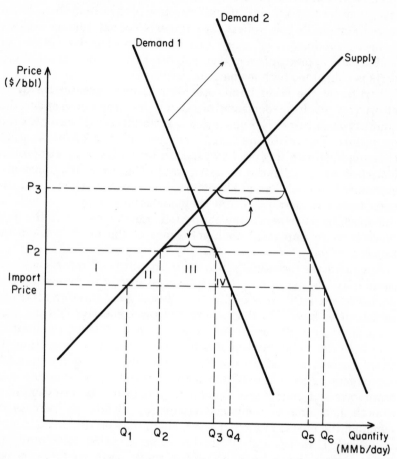

increase in imports from the original amount $Q_2$–$Q_3$ to the new amount $Q_2$–$Q_5$. Total domestic consumption increases to $Q_5$, but domestic production remains unchanged at $Q_2$. There is no additional incentive to domestic production without an increase in the price. As a result, the areas given by I, II, and IV remain unchanged. That is, there is no additional transfer of income from consumers to producers, no change in resource costs, and no change in the amount of consumption lost to consumers. Only area III, tariff revenues to the government, increases to include the additional revenues from the larger volume of imports. Per unit consumer opportunity costs do not change; the total amount of consumer opportunity costs increases only with the larger tariff revenues received by the government. This, of course, constitutes a transfer of income from petroleum consumers as a group to taxpayers as a group.

If a quota system is used instead, the quantity of imports must remain at $Q_2$–$Q_3$ after the increase in demand. This is shown in the figure by increasing the domestic price from $P_2$ to $P_3$, where the difference between the quantity of oil produced and consumed in the United States remains the same as before. In other words, if the quantity of imports is fixed, the increase in demand must be met through domestic sources, and this requires an increase in the domestic price. The price must rise enough to generate additional supply or reduce the quantity demanded until the market is cleared. Both per unit and total opportunity costs to consumers increase with the quota. Resource costs (area II) must increase to generate additional domestic supplies, and this means that the income received by existing production (area I) must increase as well. The scarcity value of imports (area III) increases as the gap between the domestic price and the import price widens for the same amount of imports. Area IV increases as domestic consumers are forced to reduce the quantity of purchases relative to the amount desired at the free trade price.

A similar set of conclusions follows if the import price should fall. With a tariff, the domestic price would fall by the same amount, leaving opportunity costs unchanged except for an increase in government revenues. With a quota, the domestic price remains unchanged and the opportunity costs increase in all four categories. In other words, the tariff permits the domestic economy to take advantage of a reduction in the price of foreign oil, while the quota does not. For domestic producers, the quota eliminates any risk of having the domestic price fall if the world price should decline. The tariff, in contrast, does not eliminate this element of risk. If the import price should rise, however, the domestic price would rise with the tariff but not with the quota. Opportunity costs remain the same with the tariff, but decrease with the quota. The quantity of imports decreases with a tariff, but remains unchanged with a quota.

Thus, the important distinctions between tariffs and quotas develop from the path of import prices over time. Preferences between them depend on goals and their relative importance. If the world price simply fluctuated, the quota would be the best device to maintain stable domestic production. If the world price rises, the quota will create smaller opportunity costs. If the world price falls, the tariff permits the domestic economy to take advantage of cheaper imports, but jeopardizes marginal domestic production. A variable tariff is sometimes suggested to handle the risk to domestic producers from fluctuating or declining prices. This form of tariff simply approximates the effects of a quota, but a quota is easier to apply.

## Discrimination

Quotas are inherently discriminatory because some companies receive the right to import while others do not. The way in which the licenses are issued necessarily benefits some and harms others. Typically, licenses are distributed on the basis of historical experience. Those importing in the past are allocated licenses for importing in the future. This method of determining future import shares discourages competition by limiting the entry of newcomers.

Any nonmarket allocation system creates a potential monopoly position that could be exploited at the expense of consumers. The tariff system, in contrast, operates through the market system. The only limitation on potential importers is the ability to sell in the domestic market at the higher price.

Both tariffs and quotas may be designed to discriminate according to the source of imports. With a quota, licenses may be based on country of origin. Similarly, the tariff schedule may incorporate a risk factor that varies according to source. For example, imports from Canada might receive unlimited licenses or pay no duty, while imports from Arab countries might be limited or pay the highest duty.[9] However, it would be difficult for the United States to discriminate among the sources of its imports, just as it was difficult for exporters to control the ultimate destination of their exports during the recent embargo. Substantial leakage may be expected in both cases due to the fungibility of oil in world markets. Oil can be transported to intermediate locations for retransport to the United States. Alternatively, oil from allowable sources may be switched to the United States only to be replaced by oil from restricted sources. Similarly, a price differential in one segment of the market cannot

---

[9]This approach violates the most-favored-nation clause of several U.S. trade agreements, but such violations are usually permitted under accompanying national security clauses.

last indefinitely. Actions of buyers and sellers eventually force any differences to disappear as prices converge.

## Administrative Flexibility

Under existing law, quotas are easier to implement and change than tariffs because quotas can be altered by executive action without formal procedures or legislation. Changes in tariff rates, except under special circumstances involving national security, require legislative approval. This greater flexibility is advantageous if shifts in market conditions warrant changes in the restrictions placed on imports. It is conceivable that flexibility could be built into the tariff schedule through provisions for administrative action in response to certain changes. Such flexibility would come at the cost of restricting public debate and representation.

The costs of administering quotas are generally higher than those of administering fixed-rate tariffs. A larger administrative apparatus is required to control the trading pattern and volume and to issue and review the forms traders are required to fill out. If tariffs were employed on a variable or differential basis, however, it is unlikely that there would be a significant difference in administrative costs between quotas and such tariffs.

## Scarcity Value of Imports

The difference between the import price and the domestic price created by import controls constitutes a scarcity rent that may accrue to the government, the importer, or the exporter. If the scarcity rent is collected as a tariff, it has less redistributional effect than the quota, since taxpayers conform more closely to consumers as a group than to producers.

Under some circumstances, the exporting countries may succeed in capturing the scarcity rents of imports. It was mentioned earlier that an exporter cartel could capture the rents if a quota system were used by simply charging the importer the higher domestic price (less delivery costs). This is not possible with a fixed tariff duty because the domestic price is based on the import price plus the duty. Raising the import price simply squeezes the amount of demand for imports. If a variable tariff is used—where the duty varies with the import price in order to maintain a stable domestic price —then the exporter cartel has the incentive to raise its selling price to the maximum domestic target price and absorb the flow of revenues that would otherwise go to the U.S. government.

## Conclusion on Tariffs versus Quotas

The inability of a quota system to respond to small changes in demand or supply of other fuels provides a compelling reason for policymakers to spurn the use of this form of import control. Past experience with quotas, as discussed briefly in chapter 1, suggests that the prospective dangers of interfering with the price mechanism are usually large and unforeseen.[10]

Because the tariff works through the price system, and because domestic prices have the opportunity to reflect changes in market conditions, the tariff is the preferred control system if one is to be used at all. Price flexibility is also a reason to prefer a fixed over a variable tariff. Finally, the tariff reflects explicitly the cost of import controls for public scrutiny and debate; the costs are not hidden as with quotas.

It is recognized that a tariff leaves domestic industry open to greater risk from cheaper imports than does a quota. Nevertheless, it is also undesirable to eliminate completely the pressures of foreign competition from the domestic market. If lower world prices do not result from competitive conditions but rather arise from an attempt by the cartel to disrupt domestic production, an additional protective response is required. Temporary antidumping controls would be preferable to ongoing quotas for this purpose.

The potentially massive costs of import controls associated with a move toward self-sufficiency raise the question of alternative ways of achieving energy security. Standby storage and shut-in capacity are among the alternatives considered in the next chapter as ways of achieving short-term security. Chapter 6 suggests that interdependence rather than independence may be a better approach to long-term security.

---

[10]For a more detailed discussion, see *Oil Imports and Energy Security: An Analysis of the Current Situation and Future Prospects*, House Ad Hoc Committee on the Domestic and International Monetary Effect of Energy and other Natural Resource Pricing, 93rd Cong. 2nd sess. (September 1974), chapter III.

# 5

# Direct Policies to Enhance Energy Security

Direct government action can help the U.S. economy achieve energy security goals. Interventionist policies to achieve energy security can be placed in four classes: preventive measures to insure against the effects of interruptions in supply, subsidies to increase domestic energy production, ancillary policies that increase energy security, and policies that reduce energy demand. Unless otherwise specified, the presentation in this chapter proceeds from the assumption that the domestic industry operates competitively in an economy open to imports.

## PROTECTION AGAINST SUPPLY INTERRUPTION

The problem of supply interruption is bounded by the likelihood of interruption and the quantity of supply that would be affected. An embargo is essentially an economic weapon used for political ends. The many political differences among OPEC members all but preclude their finding a common rallying cause or target for an embargo. Moreover, the costs of an overall OPEC embargo would be borne unequally. Venezuela, for example, exports nearly two-thirds of its output to the United States. Some countries, on the other hand, export very little to the United States. To bring all OPEC member states into an embargo would be exceedingly difficult. Consequently, action to minimize the damage from a complete OPEC embargo is not required because its likelihood is so small.

Embargo by a group of OPEC nations is a different matter. The Arab countries have embargoed the United States before, and the geopolitics of the Middle East make future use of the "oil weapon" a continuing possibility. As noted in chapter 2, U.S. dependence on Arab imports could grow to as much as 53 percent of total imports, or 29 percent of consumption, in 1985. Interruption of imports of even much smaller volumes would have serious effects on the U.S. economy.

Estimates of imports from Arab countries presented in table 2-4 range from a low of 1.8 MMb/day to a high of 7.9 MMb/day. These are the quantities at risk. The period for which they would be at risk is uncertain. While the Arab producers would be under no economic pressure to resume sales to the United States for some time (if current estimates of potential financial reserve holdings are correct), an embargo tends to lose its effectiveness the longer it is in place. Not only does time allow domestic demand and supply to adjust to an interruption, but it allows the strategies for circumventing the selective embargo to become effective. On these grounds, effective protection against the worst results of an embargo would be provided by capacity to replace Arab imports for approximately 180 days. Immediate action to reduce consumption and increase both domestic output and imports from other sources would stretch 180 days of storage to something like one year's embargo protection.

Protection against an embargo can be achieved through storage, through standby capacity, or through some combination of the two. It could also be provided, of course, through a policy of autarchy created through import controls of the sort discussed in chapter 4.

## Protection Through Storage

Storage can be maintained either underground (in salt domes or in natural or man-made caverns) or in steel tanks. Salt dome storage is the least costly of these options, but it is limited both in amount and in location. It is safe to assume that incremental storage for security purposes must rely on expansion of tankage. At an oil price of $7.50/bbl, the annual storage cost in steel tanks would run about $2.04/bbl; at an oil price of $10/bbl, about $2.29. The annual cost of storing 180 days of Arab imports at the $7.50 price would be $661 million if imports were at the 1.8 MMb/day level, and $2,901 million with 7.9 MMb/day of imports. The annual cost of the same amount of storage if the price of oil were $10/bbl would be $742 million at the

1.8 MMb/day level and \$3,756 million if imports were 7.9 MMb/day.[1]
If it were possible to buy oil at the competitive rather than the cartel
price during some period of cartel disarray, the cost of the storage
option would be lower.

## Protection Through Shut-in Capacity

Shut-in capacity is an alternative to storage as a protection
against supply interruption. Such capacity could be developed by
prorationing existing private production or by developing new reservoirs. The cost of inducing private holders to delay production of
their reserves can be approximated as the return on the revenues
they forgo. The opportunity cost of government-owned capacity
would be the same. Because shut-in capacity implies the ability to
produce for some time, it would provide long-run emergency capacity rather than the 180 days implied in the storage option. That is,
standby capacity could be produced for the life of the reservoir, not
just until the stored oil was consumed.

Reserves to support standby capacity must be over and above the
level of reserves maintained for domestic production. The production rate of most reservoirs allows no more than one-tenth of the
proved reserves to be produced in the first year. Those capable of a
higher production rate would be prime candidates for commitment
to the standby program, and would have lower costs. To be conservative, however, the one-tenth production pattern can be assumed; 1.8
MMb/day shut-in capacity thus implies a needed increase in reserves of 6.57 billion barrels. To have 7.9 MMb/day standby requires
28.8 billion barrels. Proved oil reserves in the United States now are
approximately 34 billion barrels. The exploration effort required to
use shut-in capacity as a primary defense against import interruptions is obviously staggering, but is certainly smaller than the effort
needed to create continuing self-sufficiency. Reserves held on
standby need only be discovered once; reserves used must continually be replaced.

To calculate the cost of shut-in capacity, let $Q$ be the quantity of
first-year production to be shut in; $k$, the maximum decline rate over

---

[1]Storage costs were calculated by updating National Petroleum Council figures
(*Emergency Preparedness for Interruption of Petroleum Imports in the United States*,
July 1973), and amortizing tankage over 15 years at a 10 percent discount rate.

the life of the reserve base; $r$, the interest rate; and $P$ the price of crude oil. Then the present value of the oil asset is given by

$$V = \int_{0}^{\infty} PQe^{-(r+k)t}dt = \frac{PQ}{(r+k)}$$

for given values of $P$, $Q$, $r$, and $k$. The present value of this stream of revenues delayed 1 year is $V/(1+r)$, so the revenue lost by delaying production is

$$V - \frac{V}{1+r} = V\left(\frac{r}{1+r}\right)$$

Suppose $P$ equals $10, $r$ equals 0.10, and $k$ equals 0.10.[2] Then protection against an interruption of imports of 1.8 MMb/day (i.e., 657 million barrels per year) produces a loss of revenues of $2,986 million for the 1-year delay. The necessary payment to reservoir owners to induce the delay will be somewhat less than total lost revenues because actual production incurs additional costs. Providing protection against an interruption of 7.9 MMb/day (i.e., 2,884 million barrels per year) produces a loss of revenues of $13,109 million per year on the same assumptions.

Standby capacity would have a number of advantages. First, it would preserve the ability of U.S. consumers to take advantage of lower world prices that might occur. The issue of energy security would not prejudice the allocation of resources. Second, there would be an advantage in retaining unutilized petroleum reserves until the dimensions of new energy technology became clearer. In essence, the nature of wisdom in the allocation of depleting resources over time is to use them up while alternatives become practical. If we retain a portion of our oil reserves and use those of other countries instead, the uncertainty in developing new energy technology would be lessened for this country. Third, there would be clear environmental benefits in retaining reserves underground rather than in tankage. While environmental costs are included in the costs of

---

[2]The $7.50 price appears irrelevant. Development of the reserves necessary to achieve freedom from Arab imports would require exploration beyond the level at the $7.88 price. The required price in the worst case would not be above the approximate self-sufficiency price of about $10.

storage, the possibility of an unanticipated cataclysmic disruption, or of unacknowledged marginal but cumulative environmental degradation through hydrocarbon evaporation, is eliminated. Finally, there is some reason to believe that use of the standby alternative would reduce the probability that OPEC would seek to use the oil weapon for political ends. The fact that standby capacity has no certain time limit, as has storage, means that an embargo implies a commitment to a virtually open-ended revenue drain for the exporting countries. They could not predict the point in time at which the United States would be forced to retreat from its position, and hence they might be less likely to initiate an embargo in the first instance.

The higher costs and inflexibility of standby capacity make it unacceptable as a total substitute for emergency storage. The benefits described above may, however, make standby capacity attractive as an adjunct to a storage program.

*Protection by Combined Storage and Standby Capacity*

A policy of combining storage and standby capacity in a single security program would probably be optimal as a means of dealing with the uncertainty of foreign energy supply. Standby capacity, in the most favorable of circumstances, cannot be made available immediately. Moreover, the process of activating fields for temporary production is expensive. There remains, too, the political difficulty of actually retaining large installations in disuse; that difficulty escalates as the size of the installation increases. Storage combined with standby capacity would significantly lower these costs.

One possible mixture of storage and standby capacity would include a 90-day additional inventory to cover Arab imports. Other combinations could be selected and costed out using perhaps other assumptions. If Arab imports were 1.8 MMb/day, the annual storage cost of this amount for 90 days would be $330 million per year at the $7.50 price, or $371 million based on a $10/bbl price. At the upper bound level of imports (7.9 MMb/day), the cost would be $1,450 million for $7.50 oil and $1,628 million for $10 oil. If Arab oil were being imported at the 1.8 MMb/day level, standby capacity of 0.5 MMb/day would be adequate with 90-day storage. The daily available crude over 180 days would be 1.4 MMb/day, including drawdown of the stored oil. This assumes that standby capacity would be put on-stream **immediately**. Since this is not likely, 1.4 MMb/day overstates somewhat the actual oil available for emergency use. The

TABLE 5-1. Storage and Standby Capacity Costs for Energy Security[a]

($ millions)

|  | Midpoint range 1.8 MMb/day insecure imports | Upper-bound range 7.9 MMb/day insecure imports |
|---|---|---|
| Storage alone, 180 days |  |  |
| $ 7.50 oil | 661 | 2,901 |
| $10.00 oil | 742 | 3,756 |
| Standby alone | 2,986 | 13,109 |
| Combination policy |  |  |
| Storage 90 days |  |  |
| $ 7.50 oil | 330 | 1,450 |
| $10.00 oil | 371 | 1,628 |
| Standby 0.5 MMb/day | 829 |  |
| Standby 2.5 MMb/day |  | 4,148 |
| Combination policy total cost | 1,159–1,200 | 5,598–5,776 |

[a]Calculated as described in the text. Standby capacity cost estimates are subject to error because of uncertainties in arriving at continuing costs.

cost of 0.5 MMb/day standby capacity would be $829 million per year on the above assumptions. The total cost of a policy combining standby and storage at the 1.8 MMb/day Arab import level would thus range between about $1,159 million and $1,200 million per year.

Standby for the 7.9 MMb/day rate of imports would be adequate at 2.5 MMb/day if there were 90 days of storage. Standby would cost $4,148 million per year at this rate. This would give a 6-month drawdown of 6.45 MMb/day, again assuming that standby is placed on-stream immediately. The cost of this package of protection would be between $5,598 million and $5,776 million. Table 5-1 summarizes these findings.

## Implementation of Storage and Standby Capacity Programs

Storage to meet the goals posited above could be accomplished by requiring importers to maintain inventories at the levels chosen. To increase imports, the importer would be required to increase his storage accordingly. Policies of this type have long been utilized in Europe. No serious administrative difficulties have arisen.

Standby capacity presents more administrative problems. The Naval Petroleum Reserves provide an institutional model of one method by which energy security can be achieved. Government agencies could be established to oversee the development of standby capacity on federal lands. Private producers in unutilized fields could be paid to maintain reserves in readiness to produce. Perhaps the purchase and mothballing of producing reservoirs would be efficient. Whatever the method used, the creation of standby reserves would require greatly expanded exploration and development of petroleum reserves. It should be emphasized, however, that over time this increase would be far less than would be required by import controls that would lead to an equal level of self-sufficiency. Consequently, this alternative is less onerous and will create fewer stresses on domestic resources than an increase in domestic production to meet all needs. The cost of standby capacity could be covered most properly by a tax on the consumption of oil, the insecure fuel.[3] The oil consumption tax would not only provide for energy security directly but would also reduce the size of the import gap because of its effects on consumption.

The administrative complexity, the political difficulty, and the inherent delays in implementation of a storage and standby capacity program should not be minimized. Broad public debate of its merits will be required. It is a policy for the long run, and cannot by itself lower the uncertainties of the next few years. Other policies more timely in their effect are considered in the following sections.

## SUBSIDIES FOR ENERGY PRODUCTION

The government may absorb directly some of the risk associated with energy production or it may subsidize such production. Each of the subsidy approaches has a number of variants with different characteristics. Before they are considered, three points should be made. First, subsidies can be conveyed directly through the appropriation and expenditure process or indirectly through tax benefits, credit guarantees, and the transfer of publicly owned resources at less than market value. Whatever the method, the result is the

---

[3]Note that importation of the other potentially insecure energy source, natural gas, would be minimal if, as has been assumed in this analysis, the field price of natural gas were deregulated.

same: taxpayers as a whole are made worse off and energy consumers are made better off. Second, subsidies are against the public interest, narrowly defined, unless (a) externalities are present which can only be exploited by collective action or (b) the income redistribution effect is both desired and efficient. Finally, the effect of a subsidy is to increase the consumption of the subsidized good, which is scarcely a desirable result when applied to energy if the size of the gap between consumption and domestic production is an indicator of energy insecurity.

The wisdom of any particular subsidy program depends on second-order effects which are often not clearly understood, and on the cost of achieving the stated goal when compared with the use of other instruments. Factors affecting the relative costs of different subsidy programs are noted in the discussion that follows.

### Government Absorption of Energy Production Risk

Producers in the United States must be prepared to compete with imported oil most likely priced between $7.50 and $10, but perhaps as low as $3/bbl, according to the analysis presented in chapter 3. Faced with these potential price fluctuations, producers would undertake a lower level of investment at any existing price than they would if that price were secure. A risk premium to cover the cost of uncertainty would be necessary to achieve the amount of output that would follow automatically from a secure price at the same level. Producers also face costs, extra but unknown, from technological uncertainty if they seek to expand energy output from unconventional sources or unknown regions. Energy supply increases when government absorbs either price or technological risk. Absorption of that risk constitutes a subsidy to energy producers. Encouraging technological change may produce beneficial side effects beyond those of greater output.

The cost to the taxpayer of a given government-supported advance in the technology of energy supply is equal to the value of the resources absorbed.[4] The resources absorbed by private expansion of technology by the same amount would presumably be larger, because technological progress has the character of a semipublic good. For the same reason, a greater transfer of income from consumers

---

[4] A discussion of risk absorption in improving energy consumption technology would be symmetrical to that presented here.

than from taxpayers would be required to achieve the same level of technological improvements.[5]

The effect of government absorption of risk is to increase the quantity of energy supplied at each prospective price. In essence, the range of possible returns below the expected value is reduced. Risk absorption reduces the transfer price of a given quantity of incremental energy without reducing the resource costs which must be covered by suppliers, and perhaps without placing more than a contingent burden on taxpayers. In achieving this effect, the government can absorb some of the risk associated with expanding energy supply in four ways:

1. By supporting research and development for new energy processes or in new producing regions.
2. By becoming the prime contractor for incremental energy supplies.
3. By guaranteeing a price floor for a given quantity of energy from specified sources.
4. By establishing a general energy price floor.

Each of these options is discussed below.

Perhaps the strongest case for government risk absorption can be made for government-sponsored energy research and development. Such activity also introduces a subsidy of the usual type in the sense that government contribution of resources reduces the cost to the private sector. The risk component alone is considered here; the subsidy component of government research and development is subsumed under the capital subsidy discussed below.

Procedurally, a government agency in pursuit of risk absorption could catalog unproven prospective energy supply techniques, array them in terms of their probable cost-effectiveness, and fund those necessary to meet the appropriate quantitative energy goals. Each prospective technique would be pursued until it was proven to be technically and economically feasible or infeasible and then abandoned if infeasible or released for commercial exploitation if feasible. Information and patents would be in the public domain.

---

[5]The patent laws work imperfectly, and hence a firm may make a technological gain which flows in part to others without full compensation. Hence a private firm cannot afford to undertake research unless the expected value to the firm, both direct and from licensing fees, is greater than the expected cost of the development. If the firm could receive all of the benefits from successful innovation, it would undertake the project if the expected benefits to *everyone* exceeded the expected costs to itself.

There are several areas in which government research and development activity might be expanded, including a major thrust in geological activity and in development of first-generation coal and oil shale conversion plants of different types.[6] Government provision of geological data would reduce the resource cost of energy supply expansion as well as the risks. Duplication of seismic effort would be minimized; surveys of ignored and marginally promising areas could be justified; a broad range of drillers could have access to the information (and hence competition would be enhanced); and the quality of decisions on prospective drilling sites would be improved. Public development and construction of prototype synthetic petroleum and natural gas plants would facilitate simultaneous investigation of a number of alternative processes. Technological uncertainty for second-generation plants would be reduced, and rewarding processes might surface that private industry, unable to spread the costs of possible failure, would never have pursued.

There are several problems inherent in a policy of subsidizing research and development. First, it is difficult to decide the proper level of funding. Second, the avenues to pursue must be chosen in the absence of the discipline of profit responsibility. Finally, energy research and development cannot alone provide energy security.

Going beyond research and development efforts, government could absorb the risk of producing additional energy by direct involvement in production.[7] Extensive discussion of government production lies beyond the scope of this paper, but it should be noted that government production would lower real energy costs only to the extent that economies of scale are exploited, and most such economies are thought to exist at the research and development stage. Moreover, government production might well supplant rather than supplement private production, hence the effect of government entry on total output is unknown.

Another approach to government risk absorption would be through a price guarantee for fuel produced from specific grass-roots facilities. The producer would know the minimum price that he could expect from fuel produced, but he would not be protected

---

[6]We make no judgment here that these are the two most cost-effective programs. Certainly the government research and development effort in nuclear power, the major previous recipient of government energy funds, faced no such test.

[7]The institutional device selected for this function is immaterial. For example, cost-plus contracts would have the same economic implications as would a government corporation.

against unexpectedly high production costs. To fit under such a program, projects should be self-limiting as to output and should not replace projects that would be built anyway. The essence of this procedure is to select projects that are submarginal at the predicted price, but which would be economic at that price if it were risk-free. The agency in charge would need to make a large number of difficult decisions under this policy, many of which would inevitably prove wrong, and some of which could be embarrassing.

The effect of a price guarantee program can be seen in the development of synthetic natural gas (SNG) plants and in creation of liquid natural gas (LNG) import facilities. Such plants have been planned by public utilities. Public utilities are allowed to include the capital cost of facilities in their rate base and to cover the return on that investment (plus operating costs) by the rates they charge consumers. If market conditions are such that they can charge a cost-covering price (and this appears assured so long as these high-cost facilities are a small part of their total investment), the firms are relatively secure from market risk. The result has been that public utilities have pushed ahead with SNG and LNG facilities despite the fact that the free market price for natural gas was expected to be lower than its synthetic or imported equivalent. In the face of such price uncertainty, it is unlikely that firms without the special protection of public utility status would invest in LNG or SNG. The SNG–LNG situation is analogous to the building of a synthetic oil plant (with expected product costs many times the possible world price) when there is a guaranteed price for its output.

A generalized price floor for U.S. energy production would remove the risk of unexpectedly lower prices. A price floor would protect domestic producers against foreign competition (as would import controls) and against unanticipated domestic supply and demand shifts as well. The consequent reduction in risk would encourage additional supply by reducing the price uncertainty inherent in operating in an environment where dramatic price shifts have occurred in the past.[8] Implementation of a generalized price floor re-

---

[8]By the same token, a price floor would inform consumers that fuel costs would remain above the floor level. Because it would eliminate the possibility of lower fuel costs, a price floor would raise the expected price of energy. The higher expected price would alter consumption patterns in such a way as to shift the demand curve to the left. Not only would removing downside risk increase supply, it would also decrease demand. We will consider other methods of decreasing demand at the end of this chapter.

quires that potential producers be convinced that the guarantee will be validated. To achieve this goal it should be sufficient to place standby import controls and assurances against domestic market disarray into legislation. These latter assurances could include numerous devices, many of which have been utilized in protecting agricultural product producers against these same sorts of market risks.

Government absorption of risk increases energy supply except perhaps where government produces energy directly. The benefits of socialized research and development are clear—not only are private risks reduced, but resources are allocated more efficiently. Government research and development is appropriate even in the absence of an energy security problem. A case can also be made for price guarantees so that firms would expand output to the limit justified by a risk-free price. Neither of these policies significantly interferes with the allocation of resources. Direct government subsidies do.

*Payment of Direct Subsidies*

In chapter 2 the amount of an oil production subsidy required to achieve self-sufficiency was estimated. If $7.88 were the prevailing price in the domestic market, a price of $11.98/bbl would be required in the midpoint case ($17.36/bbl in the upper case) to stimulate enough domestic output to satisfy consumption. That is, a subsidy of between $4.10 and $9.48 per barrel would eliminate imports and thus ensure energy security. This extreme use of the subsidy device would mean that between $33 and $94 billion per year would be transferred from the taxpayers as a whole to energy consumers as a whole in order to achieve self-sufficiency at the $7.88/bbl price. The resource cost of increasing output by this amount would be much smaller because much of the increase would go to higher prices for oil that would have been produced anyway. While subsidy proposals of this magnitude are all but irrelevant policy options, use of the subsidy process to some degree is likely. Consequently, it deserves further analysis.

The general attributes of a subsidy are modified by the method utilized. The mechanisms of subsidies include both direct payments from the Treasury and a variety of indirect subsidies.[9] The cost-

---

[9]Firms could be rewarded for units of energy produced and sold, with direct payment from the government upon proof of fuel transfer to a consuming party. Indirect

effectiveness of the different subsidy devices varies because of inefficiencies forced into the decision mechanism by their implementation. While these distinctions need not be considered here, there are important differences between subsidies based on increased energy supply *capacity* and those based on energy *output*. Greater capacity would be implanted with a direct capital subsidy than with an equivalent subsidy for production. The public infusion of resources is direct and immediate with a capital subsidy; no uncertainty exists. Note, however, that a capital subsidy leaves a firm dependent on market price for its output decision. The firm has no choice but to operate so long as its variable costs are covered by the market price. Because variable costs are controllable by the ultimate stratagem of varying the rate of production, capital subsidies lead to greater fluctuations in output than equivalent production subsidies. With production subsidies, on the other hand, output will not be restricted so long as variable costs are covered by the sum of the market price and the unit subsidy. Fluctuations in output will be dampened with a per unit subsidy, though total capacity would be smaller. If the goal is to provide energy security, a subsidy on capacity can be presumed to be more cost-effective.

There are other reasons to favor a capital subsidy over a production subsidy. After the original infusion of capital, no intervention with the market mechanism is required. Individual consumers and producers act in a fashion that allocates resources efficiently. Furthermore, capital subsidies reduce one barrier to entry in the energy industry. More competition is encouraged than if output were subsidized. Finally, the capital subsidy mechanism provides an opportunity to consider social costs and benefits beyond the nexus of the market decision. Superior overall allocation can result from capital subsidies where externalities are important, as they often are in energy supply. Though the subsidy process itself is questionable as

---

subsidies can take such forms as the below-cost provision of goods and services to energy producers and reduced tax obligations. Research and development, subsidized transportation services, subsidized credit, use of the environment for waste disposal, access to public mineral resources and water come immediately to mind as transfers of public assets to the energy sector. Tax subsidy devices are numerous: rapid tax writeoffs, investment tax credits, and depletion deductions beyond the level required to compensate for the wastage of the capital asset are among those which have been used.

an effective or desirable government policy, if subsidies nonetheless arise, they should be placed on capital, not output.

## ANCILLARY POLICIES TO INCREASE ENERGY SECURITY

Ancillary policies may be used to increase energy security. There are a number of such policies, but only two important ones are discussed here: alteration in the method by which public lands (including the outer continental shelf) are opened for exploitation and the deregulation of the price of natural gas.

### Federal Land Leasing Policy

Private exploiters of energy materials on federal lands are selected at present primarily on the basis of a cash bonus bid. A shift away from the cash bonus system toward greater reliance on performance requirements or royalty differentials would lead to greater and more rapid growth in energy production capacity. The differential effects of cash and royalty bidding are the converse of those of the capital subsidy versus the production subsidy discussed above. With the higher initial capital requirements of cash bonus bidding, firms tend to develop less capacity, but to produce it more intensively than with royalty bidding. Royalty bidding encourages extensive rather than intensive development because exploration expenditures in new provinces fall relative to the cost of getting more production from existing reservoirs. Royalty or performance bidding might also enhance competition by lowering entry barriers for smaller firms. Moreover, the different time distribution of capital expenditures would lessen the need for joint ventures among energy companies. The reduced initial capital requirement might also increase capacity growth in periods of capital shortage, assuming that the government's effect on the capital market is different from that of private firms.

The Bureau of Land Management of the Department of the Interior is planning to experiment with different leasing patterns in forthcoming outer continental shelf sales. Analysis would suggest that diverse federal mineral access conditions would increase the quantity of domestic energy supplied at any given price. It would therefore increase energy security.

## Deregulation of the Natural Gas Industry

The deregulation of the natural gas industry[10] would lead to greater energy security because the energy import gap would be smaller. Higher field prices for natural gas would reduce the quantity of energy demanded. The amount of natural gas and petroleum produced would be increased.[11] The higher price for natural gas would also bring intrafuel shifts to more secure domestic fuels as expensive gas is replaced in some uses. For example, it is likely that use of natural gas in the generation of electricity would be virtually eliminated at a market-clearing field price for gas. Liquid natural gas imports would be reduced, lessening one source of energy insecurity. On the other hand, a free market for natural gas would slow development of synthetic natural gas capability, retarding the growth of domestic energy production from that source.

Faulty arguments are sometimes proffered to support the position that natural gas deregulation would have a minor impact on the quantity of gas produced and consumed. One argument is that unregulated intrastate sales allow the private gas market to clear. Deregulation would thus have an effect only on the quantity of gas supplied from regions (such as the outer continental shelf) where all gas must flow into interstate commerce. Further, the argument goes, deregulation would have an effect on demand only in the interstate market. Variants of this position suggest that the Federal Power Commission has already moved closer to a market-clearing price for gas. These arguments are incomplete. The price in the intrastate market is depressed by the inability of interstate consumers to bid for gas. Moreover, the mere announcement of deregulation would reduce future demand because interstate consumers would make capital goods adjustments to higher future gas prices. Finally, the current ambivalent regulatory situation has led to un-

---

[10]The deregulation of natural gas was *assumed* in the forecasts of petroleum import requirements and petroleum prices in chapter 2. Consequently, if gas prices are not allowed to reach market equilibrium, the energy insecurity situation will be worse than that derived in this study.

[11]Petroleum production would rise because greater drilling for gas would result in some unintended discovery of petroleum and additional production of a direct substitute for petroleum, natural gas liquids. It can be argued, of course, that there will be a shift against drilling for petroleum when the relative return to gas increases. A conclusion that this shift in mix, associated as it is with an increase in incentives for drilling overall, would lead to an absolute decrease in petroleum discovered does not seem tenable.

certainty that in itself has restricted supply. Public statements supporting field price regulation of intrastate sales have caused intrastate prices to bear a lower certainty equivalent.

## POLICIES TO REDUCE ENERGY CONSUMPTION

Energy security would be enhanced by policies designed to reduce the amount of energy consumed by restricting energy use generally or specifically. Selective restraint allows for the accomplishment of subordinate goals; general restraint requires fewer invidious decisions by political bodies. Some illustrative examples follow.

### Selective Restraint of Energy Consumption

One specific policy to reduce energy consumption would be to require that prospective purchasers of capital equipment be informed of its energy efficiency. This principle has been implemented for some capital goods, but its operation can be expanded, particularly in housing. The purchaser or leasor of housing is usually ignorant of construction details affecting energy use and thus is unable to anticipate prospective energy costs. In light of this consumer ignorance, the builder has a positive incentive to minimize investment in energy-saving capital. He competes primarily on the basis of the price of his housing units, and bears none of the cost of energy consumption. If the consumer were informed of the present-value equivalent of future energy use for residential alternatives, energy consumption would be reduced. More direct policies could also be used. Specific taxes based on prospective energy consumption could be levied on capital assets. Alternatively, taxes on energy used in specified ways, such as gasoline used in private automobiles, can reduce energy use by altering consumer purchasing patterns. Energy efficiency criteria could be established, and use of equipment that did not meet the standards banned. Subsidies for mass transit would encourage consumers to change from energy-intensive transport modes. Direct controls could be instituted prohibiting or limiting certain forms of energy consumption.

Selective policies to reduce energy consumption have costs beyond those of loss of the benefits of energy use. With the exception of improving consumer information, each such action would probably lower consumer welfare more than would a general policy. Use of specific consumption restraints is inequitable because it will favor some energy consumers over others. It is inefficient because it directs choices of inputs away from the pattern arising from the actual costs of factors. Finally, it subordinates consumer preferences to those of energy decision-makers.

## Energy Consumption Tax

A general reduction of energy consumption would follow from imposition of a tax on energy. A tax is equivalent to a price increase in reducing consumption. Unlike a price increase, however, energy tax revenues would not flow to producers. They would have no effect on the quantity of energy supplied. This characteristic is illustrated in chapter 2. Use of a consumption tax alone to eliminate imports would require a transfer from energy consumers to the national treasury of $44 billion per year in the midpoint case and of $85 billion in the largest imports case. Clearly, use of taxes alone to achieve energy security is not feasible. Nonetheless, partial reliance on an energy consumption tax is a real policy option, and it deserves attention.

Two major issues are involved in an energy consumption tax. First, should the tax be placed on the energy-consuming device (a capital tax) or on energy consumption itself? Second, should the energy tax be placed on all energy consumption or only on the consumption of energy from insecure sources? The issue of a capital tax versus a consumption tax has been addressed before in other guises (see sections on direct subsidies and federal land leasing). The distributional and the allocational effects of the two tax alternatives would be different, although they could be made to have an equivalent effect on consumption of energy. On efficiency grounds, and in terms of tax-effectiveness, the basis for levying the tax should be consumption. With reference to tax effectiveness, there would be no reason to tax energy from sources in which the United States is self-sufficient. A tax on insecure fuels would change relative prices and encourage consumption of the secure fuels. Total energy consumed would be reduced, but the quantity produced domestically would be increased.

Two final points should be made about the income distribution effects of an energy consumption tax. First, it is consistent with our national values that those who benefit from a service should pay for it. The cost of energy security is real. It seems fair that this cost should be borne by energy consumers in proportion to their benefit from energy security. In the absence of a better method of allocating costs among consumers, a unit tax on insecure energy sources seems reasonable. Consequently, to the extent that government will be required to act to enhance energy security, wise policy would indicate that those actions should be supported by an energy tax. The second point concerns the effect of an energy consumption tax on income distribution. Any such tax will be borne in part by the poor. Few studies of the distributional effects of energy taxes have been made. Even if energy taxes should prove to be regressive, however, the problem of the poor cannot be solved by avoiding an otherwise desirable tax device. The problem of income inequities should be attacked directly, not piecemeal at the cost of a rational energy policy.

## CONCLUSIONS

A number of possible direct government actions to increase energy security were treated independently in the preceding sections. Implementation of these actions, whether or not they were combined with import controls, would reduce dependence on foreign energy sources. The policy question is which method or combination of methods offers the lowest total social cost of attaining the security goal.

A number of the actions suggested above add to energy security at *negative* costs. That is, these actions would increase the well-being of the nation and increase energy security. In essence, practices have grown up which create inefficiency in resource allocation and bring energy insecurity. There are opportunities to increase energy security and lower total resource costs that have not been utilized. These include government support for energy research and development, revision of federal land leasing policies, deregulation of the natural gas industry, and revision of energy consumption patterns. This last example is particularly important. At present, consumers do not pay the full cost of the energy they consume. Subsidies to producers, including tax subsidies, hold down energy prices. Environmental costs are in part inflicted on others as environmental degradation and in part covered by taxpayers. The social costs of the consumption patterns associated with energy use —urban sprawl, for example—are not covered by the price of the fuel consumed. To the extent the private market cannot charge consumers for such costs, consumption taxes are advisable. Certainly consumers of insecure energy should bear the full costs of energy research and development and of protection against short-term interruptions.

Three direct policy options to further increase security have been outlined in this chapter: a combination of storage and shut-in capacity, production subsidies, and taxes on consumption. The storage plus shut-in capacity option is preferable to the other two. The cost of security through 90-day storage plus 0.5 MMb/day shut-in capacity has been calculated as no more than $1,200 million per year on the assumptions adopted here. Of course, this level of standby capacity and storage would not cover all interrupted imports forever. In the event of an actual interruption that lasted more than the planned-for period, additional costs would be incurred through loss of output due to energy conservation measures. In comparison, an increase in domestic output by the amount of Arab oil imports (1.8 MMb/day) with production subsidies would require expenditures of about $5.8 billion per year, using the midpoint estimates given in chapter 2. To reduce energy consumption by 1.8 MMb/day would entail a consumption tax of approximately $1.30/bbl, or increased consumer expenditures of $9.5 billion. The comparisons are even more striking when a pessimistic outlook for imports is considered. Protection against a shortfall of 7.9 MMb/day through 90 days' storage and 2.5 MMb/day of standby capacity would cost less than $5.8 billion a year, while subsidies required to generate that amount of domestic output would cost $37 billion per year, and consumer tax payments designed to reduce demand by 7.9 MMb/day would amount to about $4.60/bbl—to total $32 billion in revenues per year. Though these estimates are crude, the differences in magnitude are reasonable. Moreover, the relative advantage of storage and shut-in capacity is even greater when the ancillary effects of this option are compared with either the tax or the subsidy alternative.

# 6

# Interdependence versus Independence
as a Policy Option

The discussion of energy security in chapters 4 and 5 centered on policies designed to make the United States less dependent on foreign energy sources. In this context it was assumed that smaller energy imports necessarily meant greater energy security, and vice versa. This policy was found at the limit of self-sufficiency to have very high, if not unacceptable, costs. A variant of this policy, coupling storage and standby production with lower imports, has a lower cost than a policy of zero imports, but again leaves the policy maker with a set of unpalatable options. It is suggested here that the long-run well-being of the United States may be best served by accepting a reasonable level of dependence on foreign energy sources, while at the same time creating for the oil-exporting countries a vested interest in facilitating a continuous flow of oil and capital throughout the world. Action to create a world of nations fully cognizant of their interdependence may provide meaningful energy security for the United States at a lower cost than through an autarchic policy. At the same time, establishing these conditions would yield benefits to the community of nations, including those for whom energy independence is impossible.

It must be stressed again that energy security has two components: protection against political denial of supplies and protection against the economic threat of unanticipated price changes and reflows of funds that can destabilize domestic markets. Protection against the former through storage and standby capacity would form an integral part of an interdependence approach to energy security. The ability to engage freely in foreign and domestic policy demands no less. The general formulation of the concept of interdependence as a means to achieve security against the economic threat in oil imports is discussed in the following section. On the assumption that a move toward interdependence would be credible to the OPEC nations, an explicit recalculation of the expected world oil price range, and of projected U.S. output, consumption, and imports is presented. The revised costs of maintaining security against politically motivated embargoes are then expressed.

The basis for an interdependent approach is to exploit the self-interests of the oil-exporting nations by creating a market environment for petroleum and financial assets that serves producers as well as consumers. There are repeated references in the preceding pages to the interrelationships between U.S. energy policy and world oil conditions. In an exactly analogous way, the energy policies of the exporting countries are related to world economic conditions.

It will be recalled from chapter 2 that the import demand projections assumed a willingness on the part of the oil-exporting nations to supply import needs at a given price. Yet the willingness of the oil-exporting nations to supply current needs depends, not merely on the current price, but on expected future prices and on the range of options available for the disposition of their oil revenues. Future world prices depend, in part, on U.S. energy policy, but U.S. energy policy also depends on expected future prices. The long-run world oil price is limited on the upper end by the cost of oil substitutes, but those costs are not fixed with absolute certainty. They depend on the potential results from energy research. The extent of the research commitment, moreover, is directly related to the risks associated with the price and security of imported oil.

The resting place for the world oil price below its upper limit depends on the combination of demand and the ability of the OPEC cartel to restrict production by allocating quotas among its members. It is sometimes said that U.S. imports contribute to the demand pressure on prices and perhaps even to the political prestige of the cartel. It follows that if U.S. dependence on foreign oil is responsible in any significant degree for preserving the cartel, then dependence is costly to the United States. On the other hand, following this reasoning, if a U.S. policy of self-sufficiency would ultimately lead to the destruction of the cartel, then self-sufficiency will be costly as well. Self-sufficiency means that the United States maintains expensive domestic output while the rest of the world enjoys relatively cheap energy. Dependence means increasingly expensive and increasingly risky foreign oil supplies. Hence either course is unpalatable. The corollary to this argument is that the other oil-importing countries would be better off if the United States pursued a self-sufficiency policy and worse off if the United States pursued an open policy.

If the above line of reasoning is false, the appropriate course for U.S. energy policy need not involve these dilemmas, and other oil consumers may not be better off if the United States pursues self-sufficiency. Basic to the issue is whether OPEC is capable of operating as a cartel in the sense of limiting output and allocating production quotas among members in order to maintain prices, whatever the demand growth rate. So far, the cartel has not had to face up to these difficult decisions, so the appropriate test of its strength has not been observed. The discussion in chapter 3 nevertheless concludes that OPEC is likely to succeed unless structural changes occur in the world economic system. The basis for this conclusion is the fact that member countries possessing the capability to break the cartel, e.g., Saudi Arabia, have a low rate of time preference for income. The motivation for expanding output to meet the world's needs at competitive prices simply does not exist under present conditions; hence the cartel will prevail. The range of opportunities for investment of oil revenues is limited and the risks involved are high. Under these circumstances, Saudi Arabia and other countries similarly placed have little incentive to compete with other producing countries for an increased share of the world's market. Consequently, it makes little difference for world prices whether the United States adds to world demand or not. What is important is whether the United States can alter the incentives of the key producing countries.

Unless one is prepared to argue that the world price is responding in a competitive way to the forces of supply and demand, rather than to the deliberate policy of OPEC producers, it follows that the world price need not respond to incremental changes in U.S. demand following a move toward self-sufficiency. The cartel itself must be shaken to loosen control. The absence of political pressure from the United States, if an autarchic policy is followed, may simply make the operation of the cartel easier.

The conclusion drawn from the above considerations is that energy security may be best achieved by pursuing policies that would result in the United States remaining, as a matter of *positive policy,* an important oil importer. This conclusion does not suggest that the United States lessen its efforts to reduce energy consumption through efficient means, nor delay pressing for research and development expenditures to hasten the prospective replacement of oil and gas as base load fuels. It does suggest that total energy self-sufficiency is undesirable as a *goal,* irrespective of its cost.

## A Positive Program for Interdependence

The United States is the world's leading financial center and the major force behind international monetary developments. The American economy represents to potential foreign investors a vast array of investment opportunities of differing liquidity, risk, and return. No other country offers the same depth and breadth of investment opportunity. It makes a great difference to the oil-exporting countries whether these opportunities become more or less accessible in the future. In turn, that accessibility directly affects their willingness to produce oil.

Similarly, U.S. participation is required to facilitate the recycling of foreign exchange from the surplus oil-producing countries to the deficit oil-consuming nations. The dollar is the denominator for the bulk of international transactions, most oil transactions, and is the principal source of international liquidity. Close cooperation among central bankers, especially those of the United States, is required to accommodate exchanges of oil revenues. New institutional arrangements are doubtless required among official monetary authorities and between monetary authorities and private financial institutions. The active participation of the United States in meeting these new challenges is necessary if they are to be overcome.

A two-pronged approach seems appropriate to facilitate recycling of petrodollars: direct bilateral arrangements between the United States and countries with surplus foreign exchange, and multilateral arrangements between the major financial centers and the oil-exporting countries.

Direct bilateral arrangements cannot be patterned after past commodity trade agreements because, unlike commodities, asset transactions require ongoing communication and trust. Responsible appointed officials could act as intermediaries between foreign governments and prospective sellers of American private assets. In addition, the Treasury may be even more aggressive and imaginative in creating special U.S. government liabilities that may be attractive to the oil countries. Specific details of these arrangements must vary among oil countries because of their differing political and economic conditions. The common goal for all transactions, of course, would be to minimize the perceived risk of purchasing American assets. A major educational effort, both at home and abroad, would be required. The crucial domestic problem will be to convince the public that ownership of assets by foreign holders involves few significant economic or political dangers. The difficulty of this problem should not be underestimated, but the effort must succeed if foreign investment of an appropriate magnitude is to be achieved.

Multilateral arrangements could make use of existing international financial institutions. The International Monetary Fund (IMF) and the Bank for International Settlements (BIS) are obvious choices. These institutions could act as intermediaries between asset-buying and asset-selling countries, or they might issue their own liabilities. The variety of options is essentially the same as with bilateral arrangements, but the community of interest is broader. Assets that are not tied to a single country could be more appealing to the oil-exporting countries.

Recycling foreign exchange to the lesser developed countries is a more difficult issue. The oil-rich countries seem to prefer direct aid arrangements rather than acting through existing international institutions such as the International Bank for Reconstruction and Development (IBRD). A revised IBRD or BIS, emphasizing the recycling of petrodollars, may provide a more acceptable vehicle. It is unlikely that the United States would be in a position to relieve the lesser developed countries' foreign exchange problems directly with a program of either independence or interdependence. These countries will continue to have a foreign exchange problem unless oil prices fall substantially or OPEC countries provide substantial credits. It is suggested here that oil prices are more likely to fall and OPEC powers are more likely to extend credits if a policy of interdependence is pursued than if it is not.

The process of international trade and investment necessarily connotes interdependence. Trade in even the most important commodities need not constitute a serious threat to economic security. In fact, petroleum trade, if accompanied (as it must be) by external investment by the most important of the oil-exporting nations, *reduces* the leverage of oil as an economic and political weapon. Oil exporters become increasingly dependent upon the oil consumers. The consuming nations provide a means to accumulate wealth and to generate future income based on non-oil assets. If oil were denied to a consuming country, the counterthreat of expropriation of the assets of the exporting country would exist. More importantly, if oil exporters have investments in consuming countries, any threat to the stability or continuity of asset markets is harmful. Even if investments could be hidden and thus made safe from expropriation, the general effects of an embargo would affect all investors. Not only would an embargo reduce current income of oil exporters because of lower sales, it would threaten their accumulated wealth also.

Energy self-sufficiency, on the other hand, suggests that ownership of U.S. assets may be as unwelcome as control over oil supplies. To most of the oil-exporting countries, this attitude on the part of the United States would reinforce the traditional reluctance to channel oil earnings into nonliquid foreign assets. The incentive to increase current revenues would be dampened, and the desirability of restricting output to stay within overall cartel guidelines would be enhanced. Moreover, an energy self-sufficiency policy would reduce the importance to the United States of meeting international monetary challenges. Passive participation of the United States is the best that could be expected, but more than a posture of acquiescence is required to remove existing barriers to capital flows and to open new avenues for recycling oil revenues. The United States must lead the efforts to meet the investment needs of the oil-exporting nations if those efforts are to succeed. The United States will of necessity play that active role, and will be expected to do so, if it maintains an open policy with respect to oil imports. Recycling oil revenues would become absolutely essential to the well-being of this economy. The United States would actively seek and promote the investment of surplus oil revenues in the American economy. Most importantly, the decision-makers of other nations, especially the oil-exporting nations, will recognize that the U.S. stake in establishing orderly capital markets is too high for the United States to fail to act. The necessity for success would provide the credibility otherwise lacking in world financial affairs.

## DOMESTIC ENERGY POLICY AND INTERDEPENDENCE

Pursuing a policy of interdependence does not mean ignoring the short-term U.S. security problem or abandoning the domestic energy industry. On the contrary, the self-interest of the OPEC countries in reducing prices and making oil available becomes all the more obvious if the United States pursues efforts in these areas. The important conclusion that follows is that energy security requires a balance of power between the oil-exporting nations and the United States, and a balance of imports and domestic production in the United States. Our security is no more enhanced by total dependence than it is by total independence; neither extreme creates the conditions necessary for true security.

Various policies discussed in chapters 4 and 5 are consistent both with the Project Independence framework and with a positive policy for interdependence. Continued support for energy research and development, creation of emergency supply capability, and a standby tariff to reduce the price risk to domestic producers are among the preferred alternatives. Research and development, in addition to its benefits in reducing the risk involved in backstop energy sources at the end of the petroleum era, places pressure on the OPEC cartel to reduce prices. It does so, first, by lowering the self-sufficiency price in the United States and, second, by reducing the expected time span over which oil reserves may have economic value. An effective research and development program which at the same time does not have as its apparent object the premature and imposed replacement of petroleum would stabilize world energy markets. Direct investment of the funds of oil-exporting nations in such projects may contribute to the goals of interdependence and energy security. Nevertheless, the possibility of short-term breakdown in the interdependence suggested here cannot be ignored. Consequently, a program of storage and standby capacity adequate to cover U.S. needs in such a circumstance would be required. The possibility of a higher level of energy consumption and imports associated with interdependence enhances the need for a security program. The possible dimensions of the changes are noted below.

As the above discussion implies, interdependence *does* mean a relationship that flows both ways. Oil-exporting nations and consuming nations alike would have a common interest in preserving a system of international trade in petroleum and capital. That common interest would not be based on good will (although friendly relations may help), but on the vital needs of individual nations that have a stake in making the system work. Other oil-consuming nations have little choice but to depend on foreign oil supplies. They have an even greater stake in securing the benefits of interdependence between consuming countries and producing countries. If the United States takes the lead in this endeavor, it is very likely that other consuming nations will follow. The scramble for "special relationships" that occurred in 1973 left the oil importers and some oil exporters stung by the problems of disorderly markets and unconsummated agreements. Political and economic problems generated by the appearance of coercion and panic left participants with the aftertaste of duress. The loss of confidence in the integrity of international trade will require some time to dispel. Against that background, a different approach to petroleum dealings may be welcome in many quarters.

Interdependence among oil-exporting and importing nations would require both time and innovative approaches. The mechanics of such an effort have not been explored in detail here, and the difficulties should not be underrated. It is basically a matter of constructing successful institutional arrangements among peoples with different interests. It is therefore regarded as feasible; there have been successes in similar endeavors. Assuming a successful program is implemented, the possible effects can be deduced using the analysis of earlier chapters.

## THE CARTEL PRICE WITH INTERDEPENDENCE

The proper place to begin a reevaluation of the world oil market under a regimen of interdependence is to observe what changes might be anticipated in the behavior of exporting countries. (See chapter 3 and the discussion there of influences on discount rates of member countries.) There are at least three factors that would be changed if, through a positive program for interdependence, ample and safe investment opportunities would be created for oil-exporting countries. Their foreign exchange coverage of imports would fall as short-term funds were invested in earning assets. Their limited ability to absorb capital internally would become less of a constraint on oil output as foreign investment became an attractive substitute for domestic investment or for retaining oil in the ground. Finally, current consumption capacity of the oil-exporting nations would be less important as a determinant of behavior as augmented consumption in the future became an accessible option. In each of these categories, then, better external investment opportunities would lead to a higher discount rate for an oil-exporting country and thus to a greater preference for current income.

The representative 7 percent intermediate discount rate selected for countries without pressing needs for current income would in these circumstances become appropriate for the nations of Kuwait, Qatar, Saudi Arabia, and the United Arab Emirates as well. As was noted in chapter 3, the 7 percent discount rate is arbitrary. An explicit figure was needed for illustrative purposes so that the decision-making process could be clearly identified. Seven percent was selected to represent something approaching a long-run rate of return on capital. The specific rate selected is not important to this portion of the analysis; what is important is that in a context of interdependence the rate increases, with significant implications for the world price of oil.

Interdependence would not change the OPEC countries' perception of the upper bound of the cartel price. It would, however, alter the reservation price of those countries lacking domestic consumption and investment outlets for their funds.[1] The mechanism for determining the lower bound of the cartel price remains the same in a world of interdependence as that described in chapter 3. The oil-exporting nations as a group continue to face the problem of restraining output to the amount that can be marketed at the upper bound cartel price. Attempts to expand sales by those nations with a high preference for additional current revenue will continue to place pressure on the world price. That price will decline, as demand conditions warrant, until a price is reached where some important producer will either further restrain its own production or else discipline other cartel members with threats of aggressive price cutting.

---

[1] The upper bound price would continue to depend on the perceived self-sufficiency price in the United States. The ability to obtain that high a price in the future might be reduced by a policy of interdependence, however, and hence the reservation price discussed below might be eroded. At the same time, however, the narrowing of the gap between high and low reservation prices due to the higher discount rates for some countries might reduce the possibility of complete cartel dissolution. As outlined below, the interests of the OPEC countries would be brought more in concert, reducing the incentive for actions that lead to cartel breakup even while market opportunities in the United States made competitive behavior easier. In any event, lower quasi-cartel prices would result.

Table 6-1 presents a revision of the estimates of table 3-3 for four countries, based on the change in discount rate induced by interdependence. The effect of this change is to lower the reservation price from a high of $7.88 for Qatar in the independence mode, to highs of $5.53 for Kuwait and $5.25 for Saudi Arabia in the interdependence case. The range of reservation prices is cut from $5.59 to $3.24.

Again, the reservation price of oil from Kuwait and Saudi Arabia will determine the lower bound of the cartel price (see chapter 3 for discussion of this assertion). On the basis of these calculations, that price would be approximately $5.50 per barrel. Whether the cartel can achieve sufficient cohesion to sustain even these lower bound prices remains unknown. On *a priori* grounds, however, with interdependence it would appear that the enhanced stability of world economic conditions and the smaller divergence between country goals would offset the tendency to cartel disruption occasioned by the more elastic demand for exports resulting from the availability of the U.S. market. Consequently, the issue of whether the cartel would be more or less stable cannot be resolved. On the assumption that these conclusions are sound, the effects of this change in policy can be estimated for the United States.

### REVISED U.S. IMPORT PROJECTIONS AND THEIR IMPLICATIONS

A world price as low as $5.50/bbl (c.i.f. the U.S.) implies that U.S. imports would be larger than originally projected in chapter 2, assuming that the United States maintains an open import policy.[2] The domestic price would be forced down to the import price from the $7.88/bbl level. Taking the domestic supply and demand projections at the $7.88 price and applying the elasticity estimates used in chapter 2 yields an estimate of the revised domestic supply and demand projections at the lower price. These are shown in table 6-2, along with the corresponding import volume and projected Arab import share, for the midpoint case and the pessimistic upper import case.

---

[2]It is useful to consider the effect of a world price at this cartel lower bound even though nothing in the analysis argues that the price necessarily would reach or stay at this level. It provides an important reference point for policy. A higher world price, of course, implies lower import volumes.

TABLE 6-1. Comparison of Present Revenue Equivalent, Cartel Rent, and Lower Bound Cartel Prices for Selected OPEC Countries Under Independence and Interdependence[a]

| OPEC Country | Long-run cartel f.o.b. price[b] (1) | Discount rate (in percent)[c] (2) | Present revenue-equivalent[a] (3) | Cartel rent (Col. 1 - Col. 3) (4) | Lower bound c.i.f. U.S. price ($10 - Col. 4) (5) | Lower bound f.o.b. price (Col. 5 - net-back factor)[e] (6) |
|---|---|---|---|---|---|---|
| Kuwait | ($7.01) 7.01 | (3) 7 | ($4.50) 2.54 | ($2.51) 4.47 | ($7.49) 5.53 | ($4.50) 2.54 |
| Qatar | (8.24) 8.24 | (2) 7 | (6.12) 2.99 | (2.12) 5.25 | (7.88) 4.75 | (6.12) 2.99 |
| Saudi Arabia | (7.45) 7.45 | (3) 7 | (4.78) 2.70 | (2.67) 4.75 | (7.33) 5.25 | (4.78) 2.70 |
| United Arab Emirates | (8.66) 8.66 | (2) 7 | (6.43) 3.14 | (2.23) 5.52 | (7.77) 4.48 | (6.43) 3.14 |

[a]The data on a program of independence taken from table 3-3 appear in parentheses. Representative crude adjusted to 34° gravity and 0.5 percent sulfur as explained in table 3-1.

[b]From table 3-1, column 6.

[c]From table 3-2, column 8.

[d]Present value determined by applying discount rate for 15-year period.

[e]Netback factor from table 3-1, column 5.

TABLE 6-2. Estimated U.S. Oil Production, Consumption, and Imports at a $5.50/bbl Price

(MMb/day)

| | Midpoint estimate | Upper estimate |
|---|---|---|
| Production | 10.2 | 8.6 |
| Consumption | 25.6 | 31.2 |
| Imports | 15.4 | 22.6 |
| Arab share | 8.5 | 15.5 |

The new midpoint estimates resemble the original upper estimates in terms of the volume of imports and the Arab share of the U.S. market. Total imports are 15.4 MMb/day, with 8.5 MMb/day coming from Arab countries. Arab imports constitute 33 percent of total oil consumption in this case. The new upper estimates show Arab imports of 15.5 MMb/day, about the same as total imports in the midpoint case. They constitute half of total projected U.S. oil consumption.

The potential economic disruption from an Arab oil embargo is greatly increased at the lower world price. For reasons suggested above, it would be prudent to increase domestic storage and shut-in capacity in accordance with the larger import volume. Storing 180 days' supply of Arab imports in steel tanks would cost approximately $2.8 billion per year in the midpoint case and $5 billion per year in the upper case.[3] A combination of 90 days' storage and 2.5 MMb/day shut-in capacity would cover over 4 months of Arab imports at the midpoint rate, plus a declining rate of output over a longer period of time. Total cost of this package of protection would be approximately $5.5 billion per year. The cost of 90 days of storage and 4 MMb/day of shut-in capacity would be about $9 billion per year, and would again provide more than 4 months' protection against interruption if imports from insecure sources were 15.5 MMb/day.

This much protection is perhaps excessive under the circumstances surrounding interdependence. Not all Arab countries would be equally inclined to participate in an embargo if enhanced investment incentives existed, and they risked losing the advantages

---

[3]Using the cost estimates discussed in chapter 5 and $5.50/bbl oil. These figures compare with $0.66 billion in the midpoint case (1.8 MMb/day Arab imports) and $2.9 billion in the upper case (7.9 MMb/day Arab imports) at the world price of $7.50/bbl.

of mutual cooperation as well as much of their accumulated wealth. The focus of attention turns again to Saudi Arabia in this matter. The bulk of Arab oil would come from Saudi Arabia, but she would stand to lose the most by upsetting the market. The motivation to participate in an embargo would have to be based on extremely serious issues, not merely on a show of Arab unity.

Expenditures of $5.5 to $9 billion per year to protect against the harmful effects of an embargo are small when compared to the resource and consumer costs calculated in chapter 4 in connection with import controls. Storage and shut-in capacity would make it possible to consume oil at $5.50 per barrel rather than at the self-sufficiency price. The opportunity cost of denying the nation the advantage of imports if the world price were $5.50 would range from $32.9 billion in the midpoint case to $48.5 billion in the upper import case. Put another way, if a policy of interdependence were adopted, the opportunity cost of then refusing to allow imports at the $5.50 price would be $14 billion more than at the $7.50 price, as described in chapter 4. The "insurance" costs of protecting against embargo could be covered in part by storage costs passed on to consumers and/or in total by implementing an oil consumption tax. If passed forward, either storage costs or a consumption tax would reduce consumption and thus the import gap, but they would reduce the benefits from imports as well. A 12 percent consumption tax used to cover the costs of storage and standby reserves would lower import demand by more than 1.5 MMb/day in the midpoint case and would generate $5.8 billion in revenues, which is more than enough to cover the cost of storage and standby capacity required on the assumptions cited above.

The *gross* foreign exchange outflows of the United States associated with oil imports would increase with the lower price because the volume is so much larger. Using the midpoint estimate, it was originally estimated that the United States would import 7.6 MMb/day at the $7.88 price. The maximum foreign exchange outflows would approach $7/bbl, for a total outflow of $53 million per day. Imports of 15.4 MMb/day at the $5.50 price would involve a comparable foreign exchange cost of about $4.50/bbl, totaling over $69 million per day, an increase of 30 percent from the base case. The *net* balance of payments burden would be reduced, if not eliminated, by the return investment of petrodollars.

Needless to say, the lesser developed countries would benefit greatly by lower world oil prices. Their demand for oil imports is relatively less responsive than that of the developed countries to changes in oil prices. In these countries oil is used proportionately less as a final consumption good and more as a vital input in the production process, especially in the agricultural sector. The volume of foreign exchange earnings is the primary constraint on purchases. With foreign exchange earnings small and fairly stable, the price of oil dictates the volume of oil that can be purchased.

Improved conditions in the developed countries as a result of lower oil prices would also produce indirect benefits to lesser developed nations. Not only could their exports be higher, but the funds available for foreign aid would be enhanced. The greater freedom of capital flows would broaden the opportunities for productive investment in the lesser developed nations as well.

In summary, energy security for the United States need not drive us backward toward self-sufficiency. Even in the narrow sense of security through reduced reliance on imports, no need exists for the autarchic state. If the broader vision of security through mutual interdependence is achieved, there are policy options which improve U.S. security and lead to improved resource allocation throughout the world as well.

# Summary of Review Seminar

On October 3, 1974, a seminar was held to discuss the first draft of this study. This is a brief summary of points raised at that meeting, which centered around four main issues.

Issue 1: Forecasting domestic energy supply, demand, and imports.

The participants in the seminar were concerned about forecasts based upon supply and demand elasticity estimates that are subject to a high degree of uncertainty. Some persons felt the estimates used (1.0 for supply and −0.5 for demand) were too large, but had no evidence to support an alternative measure. It was suggested that more research was necessary to upgrade the statistical estimates. The suggestions that such research should emphasize the role of price in secondary and tertiary recovery and in outer continental shelf production were widely accepted. It was suggested that emphasis be placed on refining the cross-elasticities among energy alternatives and among alternative energy-using products. Another approach, not based on elasticities, was recommended by a few participants. If such an approach, using programmatic methods, for example, were successful, it could be used as a check on the results using elasticities.

There was doubt expressed that the gap between domestic supply and demand could be filled by imports. The actions of foreign suppliers might have the same effect as an import quota in raising domestic prices to a higher level than might be anticipated using the model in the paper.

The analysis in this work places great importance on the U.S. price of oil in determining the world price. The U.S. market is small relative to the total market of foreign suppliers, so the actions of other consuming nations are relatively more important to world prices. Even if the U.S. achieves

126

independence at a given price, it is unlikely that it could export enough energy to constrain the world price at that level.

Issue 2:   The future of OPEC and the world price of oil.

The participants generally agreed that the approach used in chapter 3 in analyzing the future of the OPEC cartel was novel and useful. Some disagreed with the conclusions for individual countries concerning those countries' long-run output and price goals. Nevertheless, it was agreed that other researchers could use the same framework, while substituting their own evaluations of the factors determining the appropriate discount rates for individual countries.

There was considerable discussion of the conclusion that the OPEC cartel would persist indefinitely. For some participants this meant there was reinforced need to proceed with vigorous domestic energy programs. It was noted, however, that the cartel's strength had not been tested yet and that the prospect of increasing excess production capacity implies that a test will occur soon. Nevertheless, a lower price is not inconsistent with continuing success of the cartel.

The estimated range of long-run cartel prices received considerable attention. The estimates were questioned because they are based on the U.S. self-sufficiency price. The uncertainties about the U.S. self-sufficiency price noted under issue 1, together with the prospectively less important role of the U.S. market in determining the world price, suggest that the cartel price estimates may be based on weak grounds. There was, however, no acceptable alternative offered to replace the approach in chapter 3.

It was suggested that U.S. policy with regard to Israel is perhaps too important to ignore in analyzing the future of OPEC. Other persons at the seminar disagreed on the grounds that this is a superficial issue for even some Arab countries, and certainly cannot explain the cohesiveness of non-Arab OPEC members. Concentrating on economic motives is, therefore, more fruitful.

Other participants suggested that this work underplays the importance of the international oil companies. The analysis assumes all decisions are made by the producing countries and that the companies are passive middlemen.

Issue 3: Protection against supply interruptions versus protection against monopoly price exactions.

The participants agreed that a distinction may be made between the risk of foreign supply interruptions and the problem of cartel pricing. The conclusions concerning storage and shut-in capacity designed to guard against supply interruptions were generally acceptable, but concern was expressed about the difficulty of achieving storage and shut-in capacity goals during a period of shortage. In addition, it was agreed that a high price would prevail whether the United States continued to import from a strong OPEC cartel or moved toward independence. To some participants, the balance-of-payments savings associated with independence caused them to favor that approach. Others noted that a payments burden would arise if the United States pursued independence and the world price declined below the U.S. self-sufficiency price. The idea of energy and financial interdependence was raised as an alternative course of action to escape this dilemma. Discussion of interdependence is summarized under issue 4.

The point was raised again whether the question of energy security can be dealt with properly when the focus of attention is almost exclusively centered on the United States. Some questioned whether security can be achieved for the United States alone, without consideration of security for western Europe, Japan, and other consuming nations. There was no agreement reached on the notion that U.S. energy independence enhanced the energy security of other consuming nations.

Issue 4: Energy independence versus interdependence, and alternative means to achieve either end.

The bulk of the discussion on this issue dealt with the feasibility of interdependence with the oil-exporting countries as a means to achieve energy security. Participants argued on both sides of the issue, but the majority at least initially were skeptical of the idea. It was noted that the idea is intriguing and interesting, but that the analysis was not extensive enough to permit any conclusions. In response, the authors noted that the same comment applied to the policy alternatives designed to achieve independence and that the paper sought merely to put the alternative within a specific framework. An exhaustive treatment of this policy alternative was not intended. Nevertheless, it was noted that there is little discussion of the means to achieve the goal of interdependence. In the opinion of some, the idea appeared desirable, but there was little basis on which to judge its feasibility. For some participants, the attitudes and bargaining power of OPEC members render the notion of interdependence unworkable.

Central to the notion of interdependence is recycling of oil revenues back to oil-consuming countries. This aspect received considerable attention, but with the emphasis focused on the financial problems of lesser developed countries and the more financially unstable of the developed countries. These countries are unlikely to benefit significantly from any U.S. policy option unless lower oil prices result. Return financial flows are not likely to be large enough, nor sustained long enough, to offset the plight of many oil-consuming countries.

To some participants, a policy of interdependence will supply little additional impetus to recycling revenues to the United States. Two opposing views were expressed that led to the same conclusion. On the one hand, it was felt that there are no important barriers, either real or perceived, to capital inflows into the United States at present, and that such a policy cannot, therefore, induce more than marginal shifts in incentives. On the other hand, it was argued that the perceived risk of investing in the United States is very great and cannot be altered easily by government pronouncements.

The effect of an interdependence policy on the discount rates of oil-producing countries was questioned. After some discussion it was agreed that the precise changes in discount rates presented in chapter 6 were less certain than the direction of change and the relative structure of rates. To some, the policy of interdependence would not have a profound effect; to others, it would have a more profound effect on individual countries than on OPEC as a whole.

# Participants in the RFF–NSF Conference on
# "U.S. Energy Policy: Alternatives for Security"
# October 3, 1974
(with major affiliation at time of conference)

M. A. Adelman
Massachusetts Institute
of Technology

Douglas R. Bohi
Southern Illinois University

Thomas Brock
National Science Foundation

Robert Burch
Rocky Mountain Petroleum
Economics Institute

Ali Cambel
National Science Foundation

Nina Cornell
The Brookings Institution

Joel Darmstadter
Resources for the Future

Warren Davis
Gulf Oil

Isaiah Frank
School of Advanced
International Studies
of the Johns Hopkins University

Edward R. Fried
The Brookings Institution

Stephen Graubard
*Daedalus*, Journal of the American
Academy of Arts and Sciences

Julius Katz
Department of State

Hans H. Landsberg
Resources for the Future

Joseph Lerner
Federal Trade Commission

Theodore Moran
The Brookings Institution

Daniel H. Newlon
National Science Foundation

Arnold Packer
Committee for
Economic Development

Thomas M. Rees,
House of Representatives
Congress of the United States

Larry Ruff
The Ford Foundation

Milton Russell
Southern Illinois University

John J. Schanz
Resources for the Future

Sam H. Schurr
Electric Power Research Institute

Joel Snow
National Science Foundation

Thomas Stauffer
Harvard University

Eleanor B. Steinberg
The Brookings Institution

Robert B. Stobaugh
Harvard University

Arlon Tussing
Senate Interior Committee

David Vance
Treasury Department

S. E. Watterson
Standard Oil of California

Robert G. Weeks
Mobil Oil

David Weil
Assistant to Congressman Rees

Fred Wells
Resources for the Future

Joseph A. Yager
The Brookings Institution